SAFETY IN PETROLEUM FACILITIES TURNAROUND MAINTENANCE

CHIDI V. EFOBI

PARTRIDGE

Copyright © 2024 by Chidi V. Efobi.

ISBN:	Hardcover	978-1-5437-8092-5
	Softcover	978-1-5437-8091-8
	eBook	978-1-5437-8090-1

All rights reserved. No part of this book may be used or reproduced by any means, graphic, electronic, or mechanical, including photocopying, recording, taping or by any information storage retrieval system without the written permission of the author except in the case of brief quotations embodied in critical articles and reviews.

Because of the dynamic nature of the Internet, any web addresses or links contained in this book may have changed since publication and may no longer be valid. The views expressed in this work are solely those of the author and do not necessarily reflect the views of the publisher, and the publisher hereby disclaims any responsibility for them.

Print information available on the last page.

To order additional copies of this book, contact
Toll Free +65 3165 7531 (Singapore)
Toll Free +60 3 3099 4412 (Malaysia)
orders.singapore@partridgepublishing.com

www.partridgepublishing.com/singapore

Dedicated To

God Almighty, for His grace and favors;

Ezioma (my wife), Chimazuru (my daughter) and Chizitere (my son) for their love, understanding, patience and encouragement.

CONTENTS

Chapter 1   Introduction 1
Chapter 2   Turnaround Safety Plan 8
Chapter 3   Selection of Contractors 14
Chapter 4   Isolation of Plant and Equipment 17
Chapter 5   Draining, Venting, Purging and Flushing 28
Chapter 6   Entry Into Confined Spaces 32
Chapter 7   Hydro Jetting 38
Chapter 8   Vacuum Truck Operations 45
Chapter 9   Abrasive Blasting 51
Chapter 10 Lifting and Hoisting 58
Chapter 11 Pressure Testing 65
Chapter 12 Electrical Safety 70
Chapter 13 Painting and Coating 75
Chapter 14 Hand and Power tools 78
Chapter 15 Cutting, Welding and Brazing 85
Chapter 16 Ionizing Radiation 89
Chapter 17 Work at Heights 92
Chapter 18 Working in Extreme Hot Temperatures 102
Chapter 19 Pre-Startup Safety Review 111
Chapter 20 Startup, Completion and Review 115

CHAPTER 1

INTRODUCTION

Turnaround Maintenance (TAM) or Turnaround and Inspection (T&I) are terms used interchangeably in the petroleum industry for scheduled and periodic shutdown (total or partial) of a plant, equipment, or facility for maintenance-related activities. These activities include cleaning, inspection, repair, refurbishment, upgrade and replacement of various equipment and vessels. This implies that the regular operations of the entire plant, parts of the plant or equipment is completely stopped to perform these activities.

Turnaround of a petroleum process plant or equipment is an unavoidable reality and common in the industry. Assets deteriorate and they are required to be periodically cleaned, checked for extent of deterioration, refurbished, and sometimes upgraded or replaced. This enables the asset to continue to be competitive and prevent unplanned breakdown.

This process poses big challenges to asset owners and these range from direct cost of maintenance, loss of production to safety of plant and people. These challenges need to be managed properly to avoid severe impact to the bottom line. This book is about managing the planning and execution such that turnaround is completed safely without loss of lives or property.

There are five phases of turnaround. Some organizations have four phases. The only difference is in nomenclature. These phases are:

- Strategic Planning or Scoping
- Detailed Planning or Preparation
- Execution
- Start-up
- Completion and Review

Strategic Planning or Scoping

The first phase in a turnaround is the scoping phase. This is when the preliminary planning is done and high-level decision taken whether or not to go ahead. At this phase, asset owners look at several indices that helps in taking those decisions.

Every organization should have some sort of vision, objectives, and strategies for achieving those objectives. At the strategic planning stage of a turnaround maintenance, the asset owners should check whether the outcome aligns with its objectives and strategies. Generally, the objective of a turnaround is maintaining asset integrity and preventing catastrophic failures with safety, financial and reputational consequences.

Each of the equipment in a plant or facility should have defined turnaround intervals, prescribed by a combination of the manufacturers' instruction, lessons learnt with similar equipment and industry best practices. The intervals should have safe ranges within which decision makers have some flexibility to make alterations, putting into consideration other indices like time frame, budget, and details of assets to be impacted.

Once decision has been taken to progress with the turnaround, the following should be done in this phase:

- Selection of key personnel that will manage the turnaround (for examples, turnaround maintenance manager, engineers, planners, contract advisors, safety coordinator, operations representative, procurement officers, environmental coordinators, security officer, public/government liaison);
- Identification of other plant activities including equipment replacements and projects that can be aligned to be carried out during the shutdown window;
- Setting out the turnaround goals and key performance indicators to be used in measuring progress;
- Ordering long lead items;
- Development of progress meeting and communication plan;
- Work execution strategy- types of jobs to be executed internally, which to contract and contracting strategies;
- Preliminary cost estimate.

Detailed Planning or Preparation

When the scope is defined, the next stage is to start detailed planning and preparation. This is normally the longest phase of a turnaround, taking between 6 – 18 months. At this phase, information gathered during the scoping phase is used to carry out detailed planning. This includes all the work that needs to be done to ensure a safe and successful turnaround execution.

The following are accomplished at this stage:

- Completion of the detailed scope of work;
- Checking that all long lead items have been ordered and confirmation that they will be received on time as required;

- Development of engineering designs, packages and conducting studies, where required;
- Materials procurement and logistics arrangements;
- Selection of contractors and procurement of contract agreements;
- Development of turnaround safety plan;
- Development of QA/QC plan;
- Review and update of operations shutdown and start-up procedures;
- Detailed cost estimate and commitments;
- Development of execution schedule and risk assessment;
- Finalization of turnaround project organization chart;
- Mobilization of contractors' resources to site.

Execution

The execution phase of turnaround starts after the plant, facility or equipment is shutdown. It is then isolated and made safe for work to be executed. At this phase, all the schedules, plans and procedures that are developed at the planning phases will be implemented. This includes executing the detailed scope of work, implementing the approved communication plan, safety plan and QA/QC plan using personnel and contractors as detailed in the turnaround organization chart. The tasks in this phase involve inspection, corrective action, testing and certifying for readiness to be put back to operation. The key objective is to return each equipment back to service better than when the turnaround started. The activities in this phase will involve:

- Equipment isolation and de-isolation;
- Use of cranes and lifting;
- Confined space entry;
- Hydro-blasting and hydro-jetting;
- Abrasive blasting;

- Work at height;
- Welding and cutting;
- Use of compressed gas cylinders;
- Use of hand tools and power tools;
- Pressure testing;
- Radiography;
- Use of electrical equipment;
- Painting and coating;
- Work in extreme hot environment.

It is important to state that good planning and preparation should eliminate surprises that could lead to unexpected downtime. Further down in this book, the hazards associated with each of the above activities and control measures will be explained to enable adequate planning.

On completion of the mechanical works, the various equipment are boxed-up or re-assembled. A joint pre-startup safety review and walkthrough inspection involving turnaround crew, contractors, operations, maintenance, engineering, inspection and safety teams is conducted. This is to ascertain that work has been completed as planned and that turnaround tools, equipment, materials, portable offices and personnel have been removed from site. At this stage, all debris and waste materials should be cleared and access ways made free. In addition, safety-critical items and emergency response equipment should be checked to ensure that they are in their places and intact. The equipment or plant will now be de-isolated and turned over to Operations for startup.

Startup

The startup phase is a transition from completion of execution to normal operation and is extremely critical. After all tasks in the scope of work have been completed, and action items arising from pre-startup safety review are satisfactorily addressed, the startup will begin. It is an operational phase but closely monitored by support organizations. There should be a startup procedure/checklist that should have been reviewed, approved and communicated as part of the safety plan for the turnaround. This procedure/checklist must be followed. At this stage, only key personnel that are essential for the startup are allowed in the plant or facility. When satisfactory parameters are met during the startup, operations can then begin a gradual process of bringing the plant, equipment or facility back to normal.

Completion and Review

At this phase, post startup action items arising from pre-startup safety review are addressed. Some of these could be removal of scaffolds, insulation, proper cleanup and housekeeping, removal all mobile and contractor equipment brought into the facility and turnaround crew demobilization. When these have been completed, review will take place. This involves reporting, system performance analysis and debriefing. Questions to be answered include, have the plans been implemented as expected? Are there things to learn from the operation? Answers to these types of questions will help to streamline the process for the future. The review phase reconciles all tasks that have been performed. These include accounts for all cost incurred from the turnaround and check backs to confirm that all work orders and purchase orders have been adequately closed. Materials are reconciled and leftovers can be returned to storage and credited for next turnaround. Lessons learnt from safety findings are also discussed and actions to effectively address them in future operations are put in place.

Some of the challenges encountered at this phase may include insufficient data or unmeasurable key performance indicators.

In-order to be able to do good job of the review phase it is very important at the planning phases to put in place SMART key performance indicators with which to measure extent of success or failure of every index.

CHAPTER 2

TURNAROUND SAFETY PLAN

Preparations of the safety plan should start at the detailed planning and preparation phase of a turnaround, as soon as the scope of work is ready. In line with a meeting and communication plan developed at the strategic planning or scoping phase, safety plan meetings will be held to develop and follow-up on the implementation of the turnaround safety plan.

The safety plan will put into consideration the activities in the scope of work and identify tasks required to be completed in-order to execute the job safely. These tasks will be in the form of plans, procedures, trainings, competence verification, information communication, resources, procurement of special tools and equipment. For each of these identified tasks, a target date of completion and action party should be defined. Some suggested safety tasks that could be considered in a good safety plan are as follows.

- **Update and communication of shutdown and startup procedures**

Shutdown and startup of equipment are some of the most critical phases in operation of a process plant. As part of preparation for turnaround, the procedures for shutdown and startup are required to be reviewed for accuracy and communicated to the concerned operation personnel. The task to ensure that these are done should be assigned to a senior operations supervisor or superintendent.

- **Development of isolation procedures**

Prior to isolation of a plant or equipment, procedures shall be developed detailing all relevant hazardous energy isolations, isolation points, sequence of implementation and all measures required to ensure that the isolation is completed safely. This is normally developed by engineering, reviewed by a multi-disciplinary team of experienced engineering, operation, maintenance and safety professionals and approved by the plant manager. This task should be completed in good time such that required isolation devices can be checked to be available or ordered.

- **Hazard identification and risk analysis**

All contractors and teams that have jobs to perform during the turnaround shall conduct hazard identification and risk analysis of their respective activities. These should be conducted after detailed scope of work is concluded and contracts awarded. Action parties for these should be the respective contractors' managers and execution crew leaders. A member of the plant safety team should be tasked to identify all work parties and ensure that their respective hazard identification and risk analysis exercises have been conducted, appropriately documented, reviewed and approved.

- **Preparation of site layout plan**

The layout of equipment, site offices/shelters, vehicle/mobile equipment routes, materials storage spaces, tool vans, muster points, emergency response equipment, and others shall be properly planned such as to minimize the risk they may constitute. This plan should be in schematic form, possibly with legends and distances specified. It is recommended that this plan be developed by the turnaround team/contractors, who understand the characteristics of equipment and materials they are bringing into the plant. It is then reviewed by plant operations, emergency response personnel and safety professionals, who understand the hazards of the plant, emergency response and safety requirements. This should be approved by the plant manager and a turnaround team member can be assigned to ensure that these are done. This task shall be completed ahead of plant shutdown before these equipment are mobilized to site.

- **Confined space entry plans**

All jobs that required to be performed inside confined spaces during the turnaround exercise shall be identified and the entry plans developed. A space is described as being confined if it has restricted means of entry or exit, is not designed for human occupancy, and contains or has potential to contain hazardous atmosphere, any known serious safety or health hazards. Some examples of confined spaces in the petroleum industry are tanks, columns, vessels, sewers, valve boxes, manholes, deep excavations, pits, turbine compartments, underground cellars, pipes and others.

- **Safety training plan**

Training and competence requirements for all jobs that will be performed shall be identified. Plans for training sessions to be delivered before or while the turnaround is ongoing shall be developed. In situations where a plant or organization does not have the facility or competence to deliver a specialized training, the required competence certifications need to be to verified before site mobilization.

- **Safety orientation plan**

Contractors and other turnaround personnel that are new to the plant are required to be identified and plan to deliver safety orientation to them developed. The contents of the orientation shall as a minimum include safety rules, site hazards, site contacts and reporting structures, emergency response procedures, incident reporting procedures and access control.

- **Pre-execution safety workshop**

Prior to start of execution of a turnaround, it is recommended to conduct a safety workshop. Attendance to this workshop should include key operations, maintenance, engineering, safety and contractors' personnel. The contents shall include lessons learnt from turnaround exercises in the specific plant and industry, expected challenges in complying with safety requirements and open discussions/suggestions on safe execution.

- **Emergency response plan and drills schedules**

All emergency scenarios that are applicable to the work activities and nearby operating facilities shall be identified. The response procedures shall be communicated during safety orientation and drill schedules prepared.

- **Fall protection plans**

Contractors and crews that have work to be performed at height shall identify such jobs and develop fall protection plans. These plans may require scaffolds, man-baskets, access ladders, full-body harness and others.

- **Inspection of equipment and PPE**

All equipment, including PPE (personal protective equipment), to be used during the turnaround should be inspected by subject matter experts to ensure they meet specifications, in good condition and safe for use. Inspection tags and certificates may be issued to ensure that only approved pieces of equipment are brought to site.

- **Waste management plan**

A lot of wastes are generated during turnaround maintenance. Some of these wastes are hazardous such as radioactive and pyrophoric materials. A plan shall be developed, detailing all wastes that will be generated and specifying how to manage them safely.

- **Pressure testing procedures**

Equipment that will be pressure tested will be identified and pressure testing procedures prepared and approved before the commencement of activities.

- **Lift plans**

Lifting activities that will be conducted during the turnaround maintenance shall be identified and plans prepared ahead of time. These plans shall specify the types/sizes of lifting equipment, competence of equipment operators and riggers, position of lifting equipment and areas to secured while activities are in progress. These will be useful in ensuring that the right equipment and personnel are gotten at the planning phase and possibly finetune schedules to avoid unnecessary delays that may arise due to lifting in adjacent work locations.

- **Management of change**

All changes shall be itemized and management of change process followed to ensure that environmental, safety and health risks are properly evaluated and controlled prior to implementing them during the turnaround.

CHAPTER 3

SELECTION OF CONTRACTORS

A contractor is an organization or worker that has been hired to do a temporary work for a company. During turnaround, there are certain jobs that plant owners might not have the resources or expertise to execute and therefore have to be done by contractors. These contractors should be selected and managed carefully such that the turnaround maintenance is executed safely.

At the detailed planning and preparation phase, all aspects of the work to be done by contractors should be identified, considering the health and safety implications of such jobs. Selection of contractors should be based on health and safety as one of the key indices. Potential contractors should be required to first go through technical bids. The technical bids should be based on availability, technical competence, reliability and safety. While sending out request for bids, potential bidders should be communicated with the technical specifications, safety requirements of the plant, risks to contractors from the plant and any risks to workers and members of the public because of contractors on site.

Contractors may then develop method statements on how they propose to carry out the job and risk assessment of the contracted work, putting into consideration the safety specifications and aforementioned risks already communicated.

The turnaround team should make independent enquiries about the competence of each bidder, considering combination of skills, experience, knowledge and track record of performance. The level of enquiry to be made will depend on the level of risk and complexity of the job.

A combination of the bidders' submission and results of independent enquiries will form the basis for technical evaluation. It is not out of place to invite bidders for meetings to clarify any issues from either side. The turnaround team and potential contractors must get together to consider any risks from each other's activities that may affect the safety of the work force or the public, clarify grey areas and agree on specific actions. In a nutshell, technical bid evaluation helps to establish bidders' competence. The turnaround team need to:

- Select bidders with safety as one of the key conditions;
- Specify safety requirements in the bid request;
- Make enquiries and get evidence on experience, safety policy and practice, training and competence, supervision arrangements and track record of performance;
- Develop list of preferred bidders- those that can be relied on and with established arrangements for safety.

Only bidders that have scaled through a pre-determined technical competence level should be invited for commercial bidding.

In addition, the commercial bid should have some clauses on financial incentive and penalties for safety performance, which should include but not limited to invitation for future contracts.

When contracts are awarded and execution is on-going, it is very important to keep accurate records of each contractor's performance and build a database from which enquiries can be made for future technical bids.

CHAPTER 4

ISOLATION OF PLANT AND EQUIPMENT

Plants and equipment designated for turnaround are required to be isolated to reduce the risk on people working on them or other equipment in close proximity. Safe isolation is conducted such as to prevent uncontrolled release of energy or hazardous materials and protect people during intrusive work activities that are typical during turnaround. Some estimates have shown that effective isolation of process equipment prevents about 120 fatalities and 50,000 injuries annually (in the United States of America alone).

The energy sources that require to be isolated include:

- Electrical;
- Mechanical;
- Hydraulic;
- Pneumatic.

Hazardous materials that have the following characteristics are required to be isolated:

- Flammable;
- Toxic or reactive;
- High temperature;
- High pressure.

Unexpected release of any of the above energy sources or hazardous materials with the above characteristics has potential for serious harm, fatalities, equipment damage and environmental damage.

Petroleum facilities have process vessels, installations, pipelines and pipework that often contain hazardous materials that may be flammable and/or toxic, high temperature and/or high pressure. Intrusive maintenance activities like are done during turnaround may result into loss of containment of such substances. In addition, these plant and equipment derive energy from various sources that are required to be controlled.

There are several potential incident scenarios that include:

- Unexpected start-up of equipment, with potential for personnel being struck or caught by moving equipment and consequences of serious injuries or fatality;
- Electric shock, Electrocution, serious injuries and death;
- Burns, arc flash and death;
- Loss of containment of hazardous materials during isolation activities, confined space entry, while working on or in close proximity of isolated equipment and during de-isolation on completion of maintenance. Unexpected release of hazardous materials (chemical and thermal energy) has consequence of personnel exposure with serious injuries and fatalities. These include exposure to materials that could be toxic, hot, under pressure, oxygen-deficient atmosphere, flammable or corrosive.

Plant and equipment are required to be safely isolated such as to eliminate the risks of these incident scenarios.

Process Isolation Methods

There are several methods of conducting process isolations, the particular method chosen will depend on the risk assessment of the isolation tasks and work to be performed on the isolated equipment.

There are four basic isolation methods that are commonly used and do not require any specialist training. They are achieved using combinations of valves, blinds, spades and blank flanges. These are single block valve, double block and bleed, blinding and pipe disconnection. Let us now discuss each of these methods.

- **Single block valve**

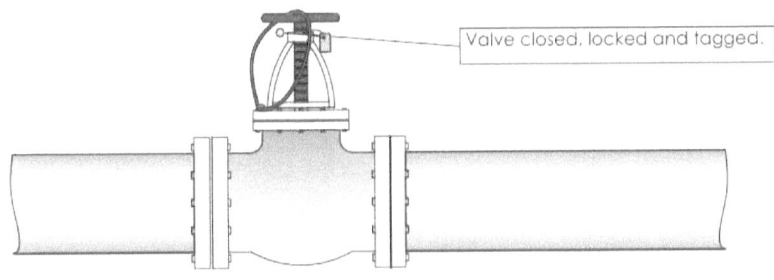

Figure 1: Single block valve

This is the simplest form of process isolation and involves closing of a block valve until the fluid is completely stopped from passing through. When the valve is closed, it is advisable to prove that the fluid has stopped passing with the use of vent, drain or pressure indicator. The valve is also required to be locked in closed position to prevent inadvertent opening and a tag installed, giving information on who closed it, when and for what purpose. The key to the lock shall be kept in a secure location that is only accessible to the individual that locked it. This method is the simplest but least desirable for hazardous materials.

- **Double block and bleed**

This method refers to closing two consecutive valves, locking/tagging them in closed positions and then opening drain or vent line between the two valves to bleed. The bleed valve is required to be locked and tagged in open position. If the bleed line is discharging, this means that either or both valves are not holding (passing). This method is more desirable than single block valve.

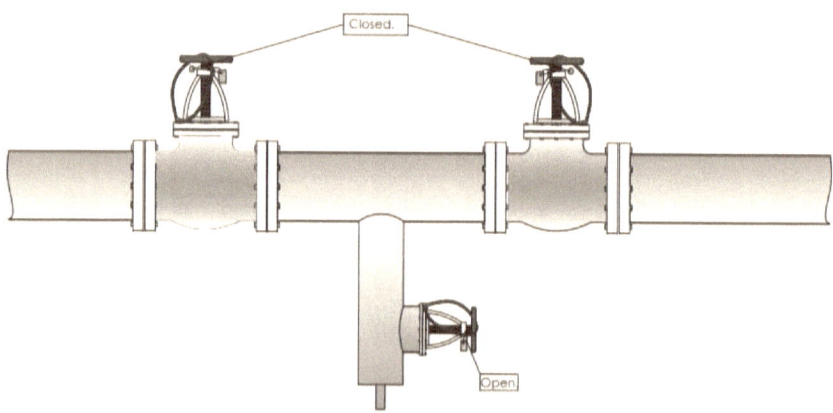

Figure 2: Double block and bleed

- **Blinding**

Blinding refers to insertion of a solid plate between flanges or at the end of pipe. The rating of a blind (designated in pounds) has to be the same with flange where it is installed. This method completely cuts off the flow of fluid. Before the installation of the blind, fluid flow is required to be stopped by the use of either single block valve or double block and bleed and proven not to be passing. In this case, single block or double block and bleed are referred to as initial or primary isolation. Blinding is a positive isolation.

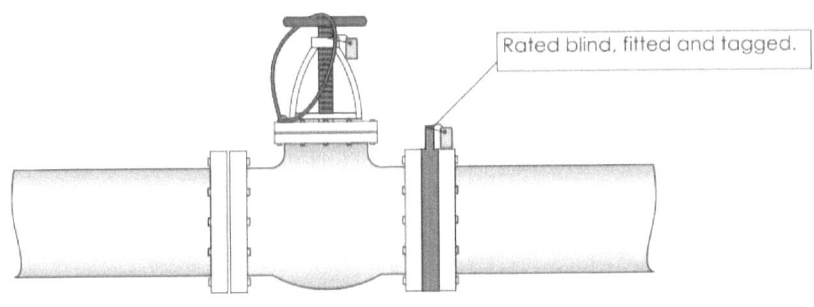

Figure 3: Blinding

- **Pipe disconnection**

Disconnection is done when the piping configuration does not allow installation of blind. Before disconnection, initial or primary isolation is also carried out and proven to be holding. Then the pipes are disconnected and blind installed at the end of the disconnected section from where hazardous materials can potentially escape. Pipe disconnection is also a positive isolation method.

Positive isolation methods are more reliable than using primary isolations only and should be mandatory on any of the following locations, work activities, characteristics of hazardous materials and work duration:

- Work to be carried out involves entry into confined spaces;
- Hot work on process piping or equipment (except steam/air/water systems less than 150 psig);

- Work on systems containing or has potential to contain flammable or toxic materials;
- Work on systems with pressure more than 150 psig;
- Work on systems with materials above their auto-ignition temperature;
- Work on systems with temperature greater than 66 degrees Celsius (150 degrees Fahrenheit);
- On the battery limits of plant or equipment undergoing turnaround;
- Estimated duration of work or isolation is beyond one shift or 12 hours (whichever comes first).

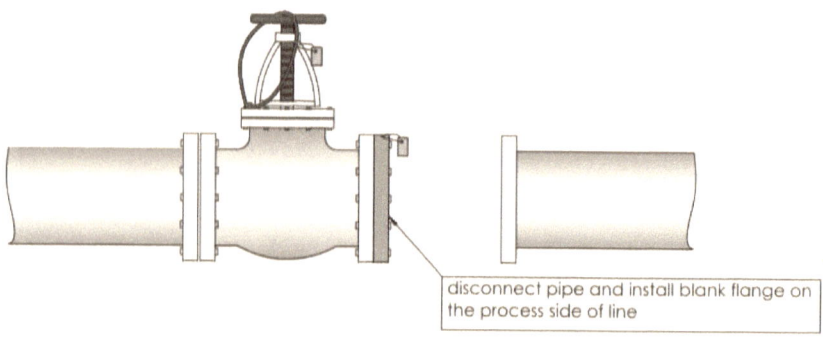

disconnect pipe and install blank flange on the process side of line

Figure 4: Pipe disconnection

There are several other specialist isolation techniques such as squeeze off, foam bagging, pipe plugs, pipe stoppers, inflatable bags, hot tapping and stoppling, pigs and pipe freezing. These are beyond the scope of this book.

Steps For Planning and Implementation of Process Isolation

The following are the key steps in planning and implementing safe process isolation.

- **Hazard identification and risk assessment**- this should be carried out on the isolation tasks and work to be performed on the isolated equipment. The energy sources, characteristics of materials involved, work activity to be performed, duration of activities and equipment to be deployed will all be put into consideration. The result of this assessment will determine the isolation methods to use.
- **Identification of isolation points** – after risk assessment, the next step is to identify the isolation points and for each point, the method of isolation. The point to isolate and the method is determined by the work location and equipment configuration, putting into consideration the result of the risk assessment conducted earlier. The rule of thumb is to isolated at the closest points possible to the work location.
- **Development of isolation procedure** – when isolation points have been identified and methods determined, the next logical thing will be to develop a procedure. This procedure should be written when there are two or more points to be isolated. The development and review should be done by a multi-disciplinary team of experienced personnel, approved by the plant manager or his designate and communicated to all personnel involved in the isolation task.
- **Installation of initial or primary isolation** – this should be done following the sequence specified in the approved isolation procedure.
- **Depressuring, draining, venting, purging and flushing** – this should be done carefully in line with the plant operational instructions and will be dealt with in another chapter of this book.

- **Testing and monitoring effectiveness of isolation** – on conclusion of depressuring, draining and/or venting, purging and/or flushing, the effectiveness of the initial isolation should be tested. If there is pressure build-up, or level begins to rise again, it implies that the isolation valves are not holding. This situation should be re-assessed by the team that developed the isolation procedure and necessary changes made. Breaking of containment lines to install blinds or disconnect pipes must not happen until this testing steps confirm that initial or primary isolations are effective. The initial isolation must be secured, monitored and maintained from this step till the intrusive work activity is concluded.
- **Line break and installation of positive isolation** – this is the most critical step in the isolation, requires necessary work permits and should be handled with utmost caution. This is done by loosening the bolts holding two flanges together, inserting a matching blind between them and closing back. In preparation for this, all hot work activities 24 meters radius of the work location should be stopped and every work group (especially those downwind) notified. Contingency plans for potential loss of containment should be put in place. If the fluid contains hazardous gases (like hydrogen sulfide), the personnel performing the work has to wear breathing apparatus until the line is broken and gas test taken to confirm absence of gas. The initial loosening of the bolt should be done at "5 o'clock position (away from personnel doing the job). When the blind has been inserted and bolts re-tightened, the work location should be re-tested to confirm that the positive isolation is effective and there is no presence of hazardous materials.
- **Intrusive work activities** – the first intrusive work activity to be carried out after isolation is completed should be closely supervised by experienced personnel.
- **Reinstatement of plant and equipment** – on conclusion of scheduled work activities, joint site inspection should

be conducted to ensure that all personnel are out, tools and materials removed and plant is safe to be reinstated. Reinstatement requires the same level of controls to those implemented during isolation. The sequence of reinstatement should be specified in a procedure. However, the common practice is to remove devices in the reserve order in which they are installed. Care must be taken when removing positive isolation as hazardous substances may build up behind the blind if the valve is leaking. It is advised to always check vents and drains before breaking the flange.

Isolation Plan or Procedure

For process isolations that involve multiple (two or more) isolation points, it is advisable to develop a written plant or equipment-specific isolation procedure. The procedure should have a drawing (preferably P&ID) marking all isolation points, sequence for installation of isolation devices, requirements to prove integrity of isolations, method of securing/monitoring the isolation devices, steps for draining/venting of hazardous materials, steps for removal of isolation devices and reinstatement. There should be a 'walk-the-plant' check against the P&ID to ensure that correct isolations have been specified and are actually in place. Also to be included in the procedure, is a list of isolation devices (valves, blinds and electrical switching devices), their positions, information on personnel that placed each device, date and time.

Non-process Isolations

Unexpected start-up or movement of machinery or release of mechanical, electrical or pressure energy has potential of serious harm of workers. The isolation of all such identified hazardous energy sources should be sequenced with process isolation.

Hydraulic, pneumatic and steam powered machinery should be isolated by closing, locking and tagging the appropriate isolation valves. Engine-driven machinery can be isolated by shutting off the fuel supply lines, locking and tagging them.

Electrical power sources should also be identified. The main power circuit and auxiliary circuits of the electrical equipment should be isolated. This can be accomplished by either switching off the circuit breakers or removal of fuses or disconnection of electrical cables or removal of plugs from socket outlets or any other means of physically preventing transmission of electrical energy to the equipment to be worked on. Noteworthy, is that whatever isolation device is used, the isolation must be secured from being inadvertently or accidentally activated. After an electrical equipment has been isolated and locked, the area should be cleared of workers and tested that there is no other way of re-starting it.

Lock-out, Tag-out

After equipment has been isolated and proven to be effective, it is important that the isolation device is secured and prevented from being unexpectedly reactivated. This is achieved by the worker (operation or electrical team member) that conducted the isolation placing a padlock on each isolation device and locking it. A tag is placed with each lock giving information on the locker, purpose of the isolation, date and time. When different personnel (maintenance and/or contractors) comes to work on an isolated equipment, he or she has to also lock and tag each of the isolation points. The keys to each of the locks are put in a secured place such that no other person can have access to them without knowledge of the personnel that placed the lock.

Turnaround activities involve hundreds of workers and it will be impossible for all to place their locks on each isolation point. In this case, it is recommended that only operations, maintenance and turnaround teams place their locks on the isolation points. The keys to these locks are then put in group lock out boxes and the boxes locked by each team. Contractors and other individuals that have work to do in the plant will subsequently place their locks on the boxes.

Figure 5: Group lock out box

CHAPTER 5

DRAINING, VENTING, PURGING AND FLUSHING

Draining, venting, purging and flushing are various methods of cleaning a vessel or equipment of hydrocarbon and other hazardous materials in-order to make them safe for work activities to be performed on them. The uses, limitations and safety issues that may arise in each of these methods are described below.

Draining

Hydrocarbon should normally be drained into a closed drain system. This will require some form of nominal pressure in the system to move the liquid into the closed drain. The drainage sequence should be specified in the shutdown procedure. However, some safety issues to be mindful of during the activity include:

- Capacity of reception facilities – reception facilities should be closely monitored to avoid overfilling or over-pressurization;

- Integrity of temporary equipment – in some cases temporary equipment like hoses may be connected to facilitate drainage of liquids or bleed off of gas. These hoses should be checked that they have embedded conductive wiring, the maximum allowable pressure is higher than that of the drain system and properly maintained;
- Formation of explosive atmospheres – gas cloud could build up due to fugitive emissions from flange connections or if the hydrocarbon is not routed to the closed drain. If unchecked, this has potential for creating explosive atmosphere;
- Asphyxiating effects of gases – gas could also have asphyxiating effects of people exposed to it;
- Static electricity – static electrical charges can build up if there is electrical connectivity, especially if temporary hoses are being used;
- Vacuum effects within the vessels being drained – vessels and tanks can compress due to vacuum effects;
- Checking that dead legs, valve cavities, valve pits and other confined portions of vessels are fully drained.

Venting and Flaring

Venting is the controlled process of depressurizing an equipment or plant by routing gas to a vent stack or to a flare header. Flaring happens when the gas is ignited in the flare stack while venting refers to release of unignited gas. Safety issues to consider when planning to vent or flare hydrocarbon gas are as follows:

- Capacity of the vent or flare facilities – the venting and flaring rates should be controlled to be within the design capacity of the vent and flare stacks;
- Uncontrolled ignition – avoiding venting when potential ignition sources exist (such as during aircraft movement, lightening);
- Formation of explosive atmosphere;

- Asphyxiating effects of gases;
- Vacuum effects within the vessels being vented;
- Static electricity – ensuring that earthing straps are fitted on equipment and structures to prevent ignition of vented gases by static electricity;
- Excessive noise.

Purging and Steaming

Purging is the injection and venting of a system with a gas or vapor to flush and displace hydrocarbon from vessels to be worked on. Important safety issues in purging are as follows:

- Design limits of the system being purged (with regards to temperature and pressure);
- Contents of the equipment;
- Layout of equipment- some vessels might have baffles that could trap gases and prevent effective purging;
- Proper control of purge rate to prevent stratification or mixing effects between purge gas and fluid being purged;
- Asphyxiating effect of purge gases – could easily displace oxygen in air;
- Formation of explosive atmosphere, if purged into the atmosphere.

Nitrogen is the most common purge gas used in the petroleum industry. The correct sequence when purging an equipment out of service is injection of nitrogen, test for any residual flammable gas and then ventilate. When purging into service, the reverse sequence applies.

When steam is used for purging it is referred to as steaming. It is useful for purging high boiling point substances and scouring the vessels and pipelines surfaces. Steaming has potential to produce static electricity and therefore good earthing is required. Particular attention should also be given to expansion and condensation effects, especially when steaming fixed-roof storage tanks. To avoid damaging the tanks, large vents must be in place.

Flushing

Flushing is the continuous injection and draining of an equipment or vessel with a liquid (like water) to clean out hazardous substances. Draining and venting could leave hazardous residues trapped in some parts of an equipment (such as instruments, dead legs, valve cavities, internal fittings, structures of floating roof tanks). These could release trapped gases when temperatures increase or the equipment is agitated. Therefore, after draining and venting, it is advisable to conduct flushing.

Water is the most commonly used flushing liquid and it is best for removing water-soluble materials but not very effective for removing hydrocarbons unless under medium or high pressure.

On the other hand, water has the potential to increase corrosion risks. In addition, equipment and piping need to be designed to withstand the weight of water.

CHAPTER 6

ENTRY INTO CONFINED SPACES

A confined space can be defined as any space or an enclosed nature:

- Large enough for a worker or workers to go inside and perform some work activities;
- Not normally intended for human occupancy;
- In which entry, movement within or exit is restricted.

Some examples of confined spaces include tanks, vessels, deep excavations, sewers, valve boxes, ductworks, pipes and combustion chambers in boilers. Once any part of the body passes through an opening into a confined space it is considered to be an entry. A number of people have been killed or seriously injured in confined spaces and these include those doing the actual work and those who try to rescue them in emergencies. US bureau of labor statistics estimates that between 2011 to 2018, a thousand and thirty (1030) workers died from occupational injuries involving confined spaces. During turnaround, workers are required to carry out several inspection and maintenance activities inside confined spaces.

Potential Hazards of Confined Space Entry

Some of the potential hazards in confined spaces are: lack of oxygen; presence of hazardous gases, fumes or vapor; presence of liquids and solids which can suddenly fill the space or release gases into it when agitated; fire and explosions from flammable vapors or excess oxygen; residues left in tanks, vessels or pipes that can give off gas, vapor or fume; limited visibility, exposure to electric shocks, falls to lower levels and high temperature. Some of these hazards may already be present in a confined space prior to personnel entry or may arise from the work activity or lack of proper isolation.

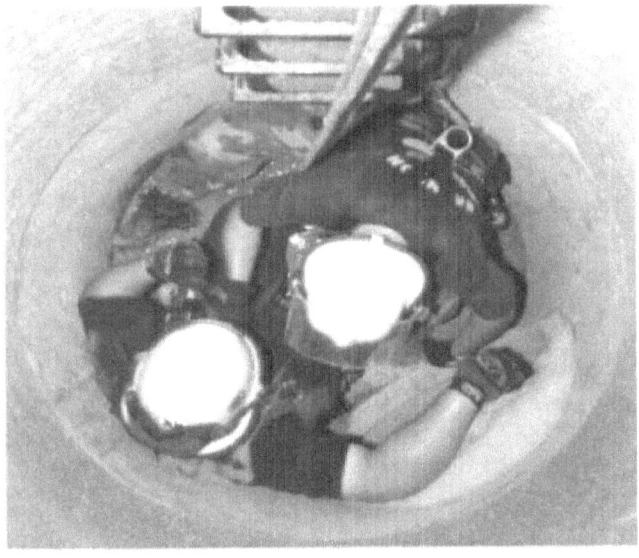

Figure 6: People working inside a confined space

Requirements for Safe Confined Space Entry

Entry and work in confined spaces should be properly planned and executed. Planning involves identifying the hazards present, assessing the risks and determining adequate precautions to take prior to entry. A complete and sufficient assessment will include consideration of the following:

- Type of task to be done in the confined space;
- Environment of the plant or equipment;
- Materials to be used in doing the work;
- Tools and equipment to be deployed;
- Fitness of personnel;
- Emergency response arrangements.

If an assessment reveals risk of serious injury or fatality, the following should be done: avoid entry to the confined space and rather do the work from outside; where entry cannot be avoided, a safe system of work should be put in place and implemented; and prior to entry, adequate emergency response arrangements should be put in place.

Safe System of Work in Confined Spaces

In a situation where entry into a confined space cannot be avoided, a safe system for working inside the space should be put in place. In order to minimize or eliminate the risks involved, the results of the risk assessment should be a good guide on the required precautions. This guide is sometimes referred as a confined space entry plan. The following includes some of the features of a safe system of work.

- Dedication of Confined Space Entry (CSE) Supervisor – CSE Supervisor should be dedicated and given the responsibility to ensure that all precautions (from the risk assessment and local requirements) are taken and should be present on site throughout the period that personnel are inside the confined space.
- Personnel Fitness to Work – The personnel engaged in the confined space entry should have sufficient experience in the type of work to be carried out and be properly trained on working in confined spaces. Their physical sizes should also be checked putting into consideration the space layout constraints. In addition, they should be medically fit for such

jobs, be fit to wear breathing apparatus and should not have claustrophobia.
- Isolation – The equipment to be worked on should be isolated, locked out and tagged out. For process isolation, the only method of isolation suitable for personnel entry into a confined space is positive isolation (blinding or pipe disconnection).
- Cleaning – the enclosure should be properly cleaned to ensure that fumes do not develop from residues while work is being carried out. Cleaning of the confined spaces involves draining, venting, purging, steaming and flushing as applicable.
- Gas testing – gas testing should be conducted prior to entry to ensure the atmosphere inside the space has sufficient oxygen and free from toxic and flammable vapors. The testing should be carried by a competent gas tester and the gas detector used should have valid calibration. Depending on the result of the risk assessment, it might be necessary to continuously monitor the atmosphere inside the space. Oxygen concentration should be at least 20%; flammable mixtures should be less than 5% of the lower explosive limit (0%, if hot work will be carried out); and hydrogen sulfide should not exceed 10 parts per million.
- Mechanical ventilation – mechanical ventilation may be required to ensure there is adequate supply of fresh air at the right temperature. This is essential where the type of work being done or the equipment used has the potential to generate gases that are hazardous or can displace oxygen. It may be imperative to install air conditioners when the ambient temperature gets as high as 43 degrees Celsius.
- Size of manhole – the size of the manhole for the entry and exit of personnel should be checked to ensure it is big enough to allow workers wearing all necessary PPEs to climb in and out with ease and for emergency rescue.
- Provision of special tools and lighting – Where flammable or potentially explosive atmosphere is present, non-sparking

tools and explosion-proof lighting should be used. In confined spaces with metallic parts, suitable precautions to prevent electric shock (like use of extra low-voltage equipment and residual current devices and adequate grounding) should be adopted.

- Use of breathing apparatus – breathing apparatus should be provided and used if the air inside the space does have insufficient oxygen (less than 20%), hazardous fumes, gases or vapors.
- Means of communication – there should be an adequate means of communication between the people inside and outside the confined space and to call for help in case of emergency.
- Standby man – a trained standby man should be positioned outside the space to keep watch and communicate with people inside. The standby man should keep a log of all personnel inside. He should call for emergency, when required. The standby man should never go into the space to carry out rescue unless he has the right equipment and has a replacement (as a standby man).
- Warning signs and access control – within a plant or facilitate, all confined spaces should be identified and signs placed at all potential entry points warning workers of the dangers and specifying entry requirements. Some of the signs may read "DANGER, CONFINED SPACES, ENTRY NOT ALLOWED". This means that entry is prohibited. If the sign reads "DANGER CONFINED SPACE, ENTRY WITH AUTHORIZATION ONLY", it implies that entry is only allowed following the plant's authorization or work permit process.
- Means of access and egress – there should be a safe means of access into and egress out of the confined space.
- Emergency response – rescue equipment including lifelines, harnesses, and hoists should be used if the confined spaces are deeper than 6 feet (1.8 meters). Emergency response

procedures should be developed, properly resourced and all personnel trained in them.

Confined Space Entry Plan

Every entry into a confined space entry should have a plan. It is important that all confined spaces that will be entered into in the course of a turnaround maintenance activities be identified and specific plans developed. This will help in planning for all the required resources and provided appropriate training to the workers. The contents of a confined space entry plan should include: scope of activities inside the confined space; inherent hazards in and around the space; procedures for elimination of the hazards; ventilation requirements; drawing (indicating the isolation points, methods of isolation, positions of ventilation equipment and approved entry/ exit points); atmospheric gas testing requirements (what types of gas test to perform and the frequency); personal protection equipment required; methods of access into the confined space; methods of access of work locations inside the confined space; and emergency response/ rescue procedures.

All the resources identified in the plan should be inspected to ensure they are adequate and safe to use. In addition, personnel that are required to enter into confined spaces, assistants and supervisor must receive appropriate trainings. The contents of the training should include: respective roles/responsibilities; hazards of the confined space and precautions; use of PPEs and emergency response procedures.

CHAPTER 7

HYDRO JETTING

Hydro jetting is often deployed during turnaround maintenance and used to clean equipment and vessels. It is most useful in removing scales/coke deposits from heat exchanges, tubes, pipelines and vessels. This is done by the application of high-pressure water generated by hydro jetting machines that are made of positive displacement water pumps driven by electrical motor or diesel engines. These machines generate pressures up to 15000 psi.

Figure 7: Hydro jetting equipment and operator

Hazards in Hydro Jetting

The pressure of water emerging out of hydo jetting equipment requires that working with or around such equipment be handled with utmost caution. Some of the hazards inherent in this type of activity are as follows.

- **High pressure water**

Exposure of people to high pressure water has potential for serious injury and may even lead to fatality. The pressure of the water emanating from a hydro jetting machine exceeds speed of 3,300 kilometers per hour and can pierce through solid materials and human body. In addition, microorganisms, air and debris may be injected into the human body with the water and could have far-reaching consequences.

- **Flying debris and machine parts**

Debris removed by hydro jetting water propelled at great speed has potential for very serious injury to eyes, skin and body parts. It can even remove an eye ball from the socket. The same applies to lose hydro-jetting equipment parts like hoses that accidental disconnect.

- **Exposure to hazardous materials**

Hydro jetting activity is meant to remove hazardous materials from process equipment and pipes. This may lead to exposure to toxic/flammable vapor atmosphere that has potential for chronic injuries.

- **Muscoskeletal injury**

Workers conducting hydro jetting activity are always in awkward positions, in confined spaces, lifting heavy tools and materials and exposed to the reaction force of the high-pressure water. These expose such workers to muscoskeletal injuries.

- **Exposure to noise**

Hydro jetting activity generates high noise level, sometimes exceeding 85 decibels and this has potential for hearing loss, if the ears are unprotected for long duration.

- **Slips, trips and falls**

The work environment is very likely to be wet, cluttered with hoses and other attachments and hydro jetting activity are sometimes conducted at heights. These could lead to slips, trips and falls.

- **Diminished vision due to misty atmosphere**

Misty atmosphere could be created by droplets of water in the air around the area that hydro jetting is being conducted. This has potential for diminished vision of both the operators and passersby.

- **Electric hazards**

Electrical equipment and connections without proper insulation and earthing in a wet environment can lead to electric shocks.

- **Ignition Source**

An internal combustion engine- driven hydro jetting pump may be an ignition source in a hydrocarbon facility.

Precautionary and Control Measures

There are several precautionary and control measures that must be in place to minimize the risk associated with hydro jetting activities. Some of the key ones are as follows.

- Operators that use high-pressure water jetting equipment should never modify the equipment or the attachments unless they are competent or have the manufacturers written approval. Only the manufacturers or their agents are authorized to tamper with or repair any part of the system.
- Hydro jetting equipment should be regularly inspected by competent people and inspection sticker posted to indicate that this has been done before being mobilized to a turnaround maintenance site.
- The pump should have a pressure gauge that should have a scale range of at least 50% above the maximum operating pressure and mounted in a position that the reading can be easily be seen by the operator.
- The hydro jetting equipment must have at least one pressure relief device that should be set at 1.2 times the maximum operating pressure. The relief device should be regularly calibrated and a tag and/or certificate issued by a competent person to that effect. The tag or certification should indicate date it is due for re-calibration. Only equipment that have relief system within a calibration window should be used on site.
- It is advisable to use high-pressure hoses recommended by an equipment manufacturer. These hoses are specifically designed and manufactured for the equipment and application. They consist of plastic liner reinforced with layers of steel. The hoses are rated for maximum allowable working pressure and such pressure is inscribed on them. The hoses must not be operated above the maximum allowable working pressure and be checked to be higher than the set pressure of the relief device. Each hose also has a service life limit. This is often influenced by age, storage conditions, temperature of operating environment, pressure and heat cycles, exposure to chemical and handling. The manufacturer should have an instruction for inspection and service life limits. However,

it is advisable to regularly inspect high-pressure hoses and remove from service any that shows signs of wear and tear.
- Whip-checks should always be installed at hose connection points to minimize risks associated with accidental disconnections.
- The various equipment components shall be rated for their intended use and parts within the high-pressure system clearly marked with maximum allowable working pressure. It is advisable during pre-use inspection to check that fitting, couplers and other components have verifiable documentations or markings indicating their rating.
- Hydro jetting equipment are operated using either a hand trigger and/or foot pedal. The control levers must be of "dead-man switch" type. This means that it has to be of continuous pressure type and will automatically stop once the pressure is released. The control levers should also have protective covering to prevent them from being accidentally activated.
- Only trained and competent personnel should be allowed to operate hydro jetting equipment or conduct hydro jetting operation. Training should include hazards and risks associated with the activities and work locations.
- Hydro jetting operators should wear the appropriate personal protective equipment (PPE) that includes:

    o Hard hats;
    o Safety shoes;
    o Face shield;
    o Ear defenders;
    o Hydro jetting suit (PVC/TST/Turtle);
    o Rubber hand gloves;
    o Any additional PPE to protect worker from exposure to chemical substances, heat and cold.

- The work area should be barricaded with warning signs and access controlled to prevent unauthorized personnel coming close to the vicinity. PVC sheets capable of withstanding the pressure of the water jet should be used to cordon off the area of the activity.
- Other work activities in the immediate vicinity of hydro jetting should be suspended, within 40 meters radius.
- It is advisable to have at least three workers involved in a hydro jetting operation: pump operator, lance operator and a helper to assist as necessary in access control, hoses management, communication between pump and lance operators, and providing relief.
- Adequate housekeeping should be maintained around the hydro jetting area.
- When work is being performed at height, adequate work platform with guard rails should be provided with easy access.
- Flame or spark arrestor should be installed on the exhaust of diesel engine driving the hydro jetting machines.
- Ground fault circuit interrupters (GFCIs) should be used for temporary wiring and when working in areas that are wet or damp.

Figure 8: Hydro jetting equipment showing pressure gauge and pressure relief devices

CHAPTER 8

VACUUM TRUCK OPERATIONS

Use of vacuum trucks is a quick, safe and efficient method of removing hydrocarbon sludges, cleaning up spills and evacuating liquids and solids from vessels and tanks during turnaround maintenance activities. However, incidents have occurred during operation of vacuum trucks with very serious consequences.

Figure 9: Rear view of a vacuum truck

Figure 10: Side view of a vacuum truck

Hazards in Operation of Vacuum Trucks

There are numerous potential hazards associated with vacuum truck operations and some of them are described below.

- Vacuum tank can be over pressured leading to explosion.
- Vacuum truck engines, exhaust unit, overheated pumps, defective electrical wiring, static electrical discharges and third-party ignition sources are potential ignition sources during operation of vacuum trucks in hydrocarbon facilities.
- There are possibilities of spills and flammable atmosphere around the operation area due to hose failure and discharge of flammable vapors from the vacuum truck, receiving container or source container.
- Workers can potentially be exposed to toxic vapors, liquids and solids.

- Discharging flammable materials in open areas have the potential to release vapors and if there is an ignition source, can result into fire and explosion.
- There is the possibility of toxic vapors that are below hazardous concentrations before handling becoming concentrated and hence hazardous at the discharge port.
- Incompatible materials may get mixed up inside the vacuum tanker that may trigger undesirable reactions.

Precautionary Measures in Vacuum Truck Operation

There are various precautionary measures that should be in place to minimize the risk associated with vacuum truck operation. The measures are segregated into truck design and equipment, maintenance, setup and operation.

Vacuum Truck Design and Equipment

- Vacuum trucks have different designs for different purposes. Those with pressure tested cargo tanks are used to collect and transfer petroleum liquids and hazardous wastes and products. Pneumatic cargo tanks are designed to collect and transport non-hazardous materials.
- Vacuum cargo tanks for flammable and combustible liquids should have the design pressure to be at least 25 psi and pressure tested to at least 40 psi. The design and test pressures should be inscribed by the manufacturer on a name plate attached to the tank shell.
- To prevent overflow and over pressure, vacuum cargo tanks are required to have level indicators and pressure gauges. In addition, there should be a calibrated relief valve or rupture disc and a means of manually depressurizing the vessel.

- It is important to use only vacuum pump that are powered by the truck engine rather than those driven by external motors.
- The vacuum pump should come with a relief device. This should be regularly calibrated and checked to ensure that it is not overdue before use.
- Diesel powered vacuum truck engines are safer to use in hydrocarbon service because of the limited electrical systems that may constitute sources of ignition. The limited electrical devices and wirings should be properly maintained to prevent lose connections that have potential for arcing.
- While working in hydrocarbon facilities, there is the potential of flammable vapors entering diesel engine air intake and causing "runaway" or "dieseling". Vacuum trucks should be equipped with manual or automatic emergency shutdown devices that close the air intake when 'dieseling' occurs. In addition, spark-arrestors should be installed in the exhaust system.
- In-order to minimize the risk posed by static electricity, conductive hoses are recommended for vacuum trucks used in evacuating flammable and combustible liquids. The hoses are thick walled and have embedded wiring. It is required to provide electrical resistivity less than or equal to 1 megohm per 100 feet.

Vacuum Trucks and Equipment Maintenance

Vacuum truck tractors, engines, vacuum pumps, cargo tanks, separators, valves, filters, hoses, nozzles, connectors, bonding and grounding cables and other attachments should undergo regular preventive maintenance.

Air tanks should be checked regularly for accumulated water or any kind of liquid. They should be drained daily of water, most especially if there is no air dryer.

Trucks electrical system should be inspected regularly to ensure there are no lose connections that will constitute ignition sources. Defective electrical systems should be repaired immediately.

Manufacturers' instructions should be followed with regards to installation, operation, pressure limitations, testing and maintenance of vacuum pumps. It is important to always ensure that pumps, bearings and associated equipment are well lubricated.

Pumps should be checked frequently for leaks, valve seating, housing cylinder wear and tear, impeller or rotor wear and tear. For belt driven pumps, the belts should be checked frequently for wear and slacking. If required, the belt should be re-tensioned to reduce friction and increase in heat generation.

Hoses are required to be inspected for cracks, leaks, exposed metal braids, kinks or wear points and regularly tested for conductivity. Nozzles, fittings and connections are required to be inspected to ensure that there is no blockage and that there are tight and conductive connections.

Pressure gauges, pressure and vacuum relief valves should be regularly calibrated, with stickers, tags or certificates provided to indicate date it is done and the next calibration due date.

Inspection, test and maintenance of cargo tanks and trucks should be done in line with manufacturers' instruction and local regulations.

Vacuum Trucks Setup and Operation

Before deployment of vacuum trucks, the following should be inspected, checked, done and documented:

- The vacuum truck is right for the application (solids, liquids or hydrocarbon);

- Cargo tank design pressure is not less than 25 psi and test pressure, 40 psi and these are inscribed on a name plate on the tank shell;
- Level indicators are functional;
- Valves operate freely and are not leaking;
- Pressure gauges, pressure and vacuum relief devices have stickers, tags or certificates showing that they are not overdue for calibration;
- Manual means for depressurizing the tank;
- External electrical devices and wiring in good condition (no lose connection or exposed naked wiring);
- Cargo tank is empty of any previously vacuumed material;
- Truck is equipped to direct discharge vapors at least 50 ft downwind or 12 ft above the truck;
- Truck equipped with spark arrestor that has been inspected recently;
- Hoses have imbedded conductive wiring;
- Hoses conductivity tested within the past one year and resistivity indicated to be less than 1 megohm per 100 ft;
- Hoses do not have any exposed metal braids, kinks or wear points;
- All accessories to be used for vacuuming made of conductive materials (nozzles, tubes, fittings);
- Vacuum truck bonded to all hoses and properly grounded;
- Gas test taken and area indicated to free of flammable atmosphere;
- Truck positioned a minimum of 25 ft upwind or crosswind of material to be vacuumed (in an open area) and 50 ft in a diked area;
- A dedicated fire extinguisher is available and kept close;
- Operator is trained and is aware of the hazards of the material being vacuumed.

CHAPTER 9
ABRASIVE BLASTING

Abrasive blasting is a process of using compressed air or water to direct a stream of an abrasive material at high velocity to clean an object or surface, to smoothen, roughen, shape the surface, or remove the surface coatings or contaminants. The abrasive materials used can be any of the following: silica sand, garnet sand, nickel slag, coal slag, glass, steel shot, steel grit, dry ice, plastic bead media, sponge and sodium bicarbonate.

Hazards in Abrasive Blasting

The hazards associated with abrasive blasting operation are as follows:

- Hazardous Dusts: Inhalation of the hazardous dust particles produced during abrasive blasting can result into respiratory problem, lungs damage and sometimes, heavy metals poisoning. Silica sand has potential to cause silicosis, lung cancer and breathing problems in exposed people. Coal slag, steel grit and garnet can cause lung damage
- High Pressure: People around the work area can be exposed to high pressure arising from the compressed air or water

being used. Loss of containment has potential for serious injuries that include particles becoming embedded in the skin, eye damage, severe cuts and burns.
- High noise level: High noise levels can cause hearing loss if workers and others in the vicinity are unprotected.
- Improper manual handling: Manual handling tasks are also involved. If done improperly it can result in strains, sprains, fractures, dislocations, bruises and overuse injuries.
- Confined Spaces: If the activity is carried out in a confined space there could be asphyxiation and death.
- Carbon monoxide: Carbon monoxide could be inhaled from oil lubricated air compressors used to supply breathing air resulting to death.
- Hand-arm vibration: Prolonged exposure to abrasive blasting vibration can damage the nervous system and result in a condition known as vibration syndrome.
- Heat over exposure: Use of the required PPE for long periods without adequate rest can result to heat over exposure.
- Explosive atmosphere: If explosive atmosphere exists at the work site the friction of the abrasive material and the surface being worked on can become ignition source.
- Static Electricity: There could be accumulation of static charges that can shock employees and cause fires and explosions.

Precautionary and Control Measures

In order to minimize the risks posed by these hazards, it is important to identify the hazards and assess the risks. Every abrasive blasting operation is unique because the surfaces, coatings, blast material, work environment and work conditions vary. Knowledgeable and competent personnel should conduct the risk assessment and put adequate controls in place. Some of the controls should be engineering and administrative, personal protective equipment (PPE), including respiratory protection, and training of workers. Engineering controls should include substitution, isolation, containment and ventilation. These are primary means of minimizing exposure to airborne hazards associated with abrasive blasting operation. Administrative controls should include use of good work procedures, and personal hygiene practices. The following control measures should be in place:

- Use of less toxic abrasive material should always be explored. For example, in some places, the use of silica sand is prohibited. There some abrasives that can be delivered with water (in the form of slurry) thereby reducing dust generation.
- Blast operators should be trained in the correct use and hazards associated with the equipment and materials they use. They should also be aware of the hazards control measures, personal hygiene practices, safe work practices and use of PPE, including respirators.
- Material Safety Data Sheets (MSDS) of the materials in use should be available and communicated to all workers.
- Barriers and curtain walls should be used to isolate the blasting operation from other workers. If possible, exhaust ventilation system should be used to capture the produced dust.
- Exhaust ventilation equipment should be used to capture dust in containment structures.
- As much as possible, abrasive blasting jobs should be scheduled when the least number of workers are on site and on less windy days in order to prevent the spread of hazardous dust.
- Operators should wear air-supplied hood, type "CE" (NIOSH-approved) while blasting and other workers in the vicinity must wear air-purifying respirators with dust filters.

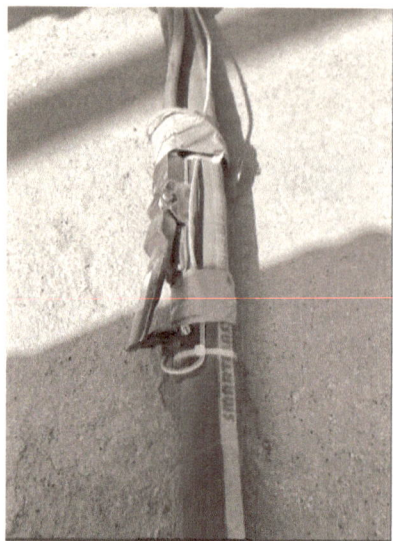

Figure 11: Abrasive blasting dead-man's switch

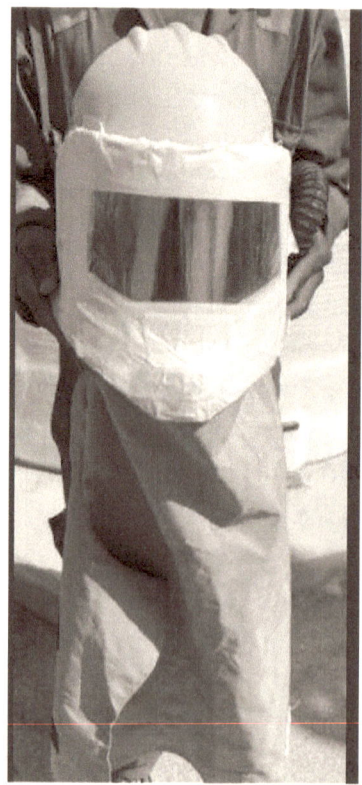

Figure 12: Abrasive blasting air-supplied hood

- Other required PPEs are hearing protection, eye and face protection, hard hat, leather gloves covering forearm, coveralls, apron, and safety shoes.
- Other workers in the vicinity of the work area should wear particulate respirators (approved by recognized national and international organizations) for airborne contaminants that are likely to be generated in the particular job.
- High pressure lines should have safety wire installed in every coupling point to prevent separation. The hoses and couplings should be inspected before use for cuts, abrasion and damage. The abrasive blasting nozzle should have a dead man switch that automatically shuts-off the flow of abrasive material if the operator releases control of the switch or nozzle
- Operators should be trained on proper manual handling techniques.
- If the activity is taking place in a confined space, all confined space entry procedures should be in place.
- Breathing air equipment in use should have high efficiency air filters, oil/water traps and in-line carbon monoxide monitoring with an audible alarm (if oil lubricated breathing air compressors are in use).

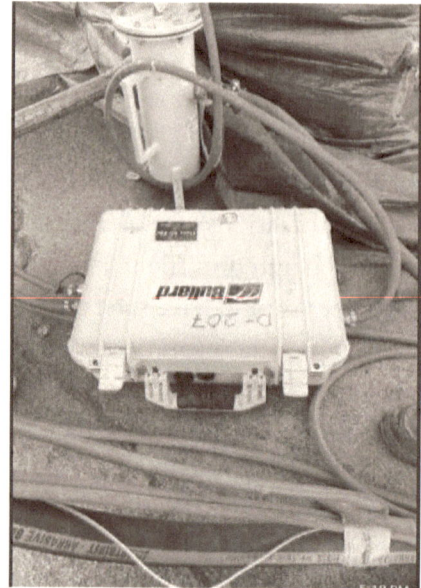

Figure 13: Carbon monoxide monitor

- Job rotations or more frequent breaks should be practiced in order to reduce the extent and duration of continuous exposure to vibration and heat.
- Gas test should be performed to ensure that flammable gases are not present.
- The blasting equipment and the structure being cleaned should be connected and effectively grounded in order to prevent static charges from accumulating.
- Warning signs and notices should be placed at strategic locations to inform other workers of the activity.
- Eye wash and safety shower stations should be provided in case of accidental exposure.
- Internal combustion engines are to be positioned downwind of the breathing air supply compressor to avoid intake of smoke.
- Eating or drinking in blasting areas should be prohibited.
- Wash stations should be provided so that workers can clean hands and face properly before eating or drinking. In addition,

abrasive blasting operators should be encouraged to vacuum or remove contaminated clothes before eating or drinking. Locker rooms should be provided where they can change out the contaminated work wears.
- Workers assigned to abrasive blasting jobs should undergo pre-placement medical examination. This examination should include chest x-ray and pulmonary function study and be repeated at least every two years.

CHAPTER 10

LIFTING AND HOISTING

Lifting and hoisting operations are one of the leading causes of fatalities and serious incidents in the petroleum industry. Every type of lift needs to be properly planned and executed subject to the requirements of local legislation in order to minimize or completely eliminate the risk they pose.

Lifting Hazards

Lifting materials during turnaround maintenance has the following hazards:

- Load falling – when a load is lifted, there is a risk of it falling and injuring people underneath or damaging property;
- Hitting and crushing – lifted loads may swing and hit nearby facilities or people;
- Crane falling – cranes, if not well positioned or loaded beyond their capacity, can fall during lifting operation with potential consequences of serious harm to people and fatalities;
- Environmental hazards – high wind speed, poor visibility due to dust or fog have potential to cause crane lift incidents;

- Damage to underground facilities – when cranes are stationed or moving heavy loads in hydrocarbon process areas, there is the risk of damage to underground facilities (e.g. buried pipelines);
- Electrical hazards – when lifting activities are executed close to power lines or electrical equipment, there is potential risk of electrocution that can lead to fatality;
- Equipment failure – if lifting equipment are not properly maintained, they can fail with potential consequences of serious incidents;
- Human error – human error such as improper use of lifting equipment or miscommunication can also lead to accidents.

Precautionary and Control Measures

It is very important that appropriate precautionary and control measures are taken when lifting and hoisting are being conducted during turnaround maintenance. These measures range from getting the right and well-maintained equipment, planning to conducting the lifting operations properly.

Lifting Equipment, Inspection and Maintenance.

- All lifting equipment (consisting of lifting appliances, rigging hardware and man-baskets) shall have their safe working load and identification number marked on them. It is advisable to only buy equipment designed to a recognized standard and fitted with necessary safety devices. These safety devices shall be operational and never over-ridden.
- All equipment shall be properly maintained in line with manufacturer's instruction, operating experience, applicable local regulations and failure modes, integrating preventive

and predictive maintenance. Auditable records of these maintenance activities shall be kept.
- Lifting equipments shall undergo thorough inspection by competent persons at a frequency not exceeding 12 months. In cases of equipment used in lifting people, the frequency shall be reduced to 6 months. Inspected equipment should be color-coded.

Figure 14: Crane

Planning the Lifting Operation

Every lift should have a plan, identifying the hazards, analyzing the risks and putting control measures in place. This can be a stand-alone document or part of other documents, the details varying with the risk and complexity of the lift. Simple and routine lifts may only require a generic plan, with an onsite risk assessment and pre-job tool box meeting. Critical lifts will require engineering design input, with drawings and calculations. A critical lift plan should be approved by a competent person.

A lift should be considered critical if:

- It is within 35 feet (10 meters) of hydrocarbon facility/equipment/piping, or pressurized equipment/piping or populated/traffic areas or railway line or the fully extended boom of the lifting equipment is within 35 feet (10 meters) of overhead power lines;
- The load is 40 tons or greater;
- The load exceeds 90% of the lifting equipment rated load capacity;
- The load has potential for explosion/fire/high heat hazard;
- It is a tandem lift (i.e. requiring two equipment to lift same load);
- The load is a man-basket;
- The lifting operation is conducted in the night;
- It is a blind lift (i.e. the load is not within sight view of the equipment operator at any point during the operation).

The lift plan should address and specify the following (as a minimum):

- The nature and weight of the load and lifting points;
- Equipment and rigging hardware required and certification checks;
- The type and number of personnel required, their roles and competencies;
- Pick up and set down points and constraints (if any);
- Step-by-step instructions;
- Means of communication;
- Emergency response plans;
- Environmental condition - weather, lighting etc.;
- Access and egress routes for slinging and un-slinging the load;

- Concurrent or nearby operations (and type of work permit, if required);
- Load integrity checks;
- Load charts;
- Assessment of necessity to use tag lines, checking any additional hazards.

Conducting the Lifting Operation

One competent person (a rigger) should be designated as the person in charge of the lifting operation and he should be at the site at all times while the activity is ongoing. The following should be his responsibilities:

- Coordinates, controls and executes the lift;
- Reviews the lift plan and ensures that each of the controls specified is in place;
- Inspects the lifting equipment and rigging hardware to ascertain they are appropriate and safe for use;
- Ensures that the equipment operator and any other person involved are competent, aware of the procedures to be followed and their responsibilities;
- Communicates the lift plan to all the people involved;
- Communicates the activity to nearby work crew and any other people that may be affected;
- Ensures the lift is carried out in line with the approved lift plan and suspends the operation if changes not envisaged in the plan takes place (for example, change in wind speed or direction).

The required competence for equipment operator, rigger and other people involved in lifting operation should be specified by local legislation.

While conducting the lifting, the following critical practices should be followed:

- Before starting lifting operations, the rigger in charge of the lift should hold pre-job meeting to communicate the lift plan to all involved;
- When lifting operations is to be controlled by hand signals, a signal man should be designated and possibly wear peculiar reflective jacket;
- The signal man should be trained and competent for the job;
- The hand signal to be used should be the universal signal understood by everybody;
- When radio communication is to be used, continuous verbal instruction shall be used and the operator should stop when there is no clearly understood signal or instruction;
- The lifting equipment operator should obey an emergency stop signal at all times, no matter who gives it;
- The lifting equipment and rigging hardware shall be rated for the load being lifted;
- The equipment maintenance records and certification should be readily available and verifiable on site;
- Lifting equipment and rigging hardware should undergo detailed examination by a competent person at least once a year (half-yearly if used to lift people);
- The lifting equipment should be inspected and documented on daily basis to ensure the safety devices are in place and functional. Defective components should be repaired before the equipment is put to use;
- Rigging hardware that have been examined and approved to be used within the year should be colored-coded. The approved color code for every period should be communicated to all personnel. For example, if in the year 2015, it could be specified that the color code is GREEN. The color code should be displayed on the notice board and communicated to all personnel. For that year 2015, all rigging hardware

that have been examined and approved for use will be color-coded GREEN (by application of the color in a conspicuous part of the hardware). However, the user should visually inspect the hardware before each use. Any defective hardware should be discarded immediately;
- Equipment operators shall be competent and licensed for the specific equipment they are operating;
- Crane cabs should have 360 degree visibility around the lift area;
- Lifting activities should not be carried out if wind speed gets over 20 miles per hour (32 km/h) or when there is poor visibility due to fog or rain;
- Lifting activities should be conducted on stable soil and out riggers fully extended on pads to avoid the crane sinking into the earth and falling over;
- Tag lines should be used to control a suspended load from swinging and hitting people or structures;
- When lifting activity is being carried out close to energized electrical equipment or power line, there should be a safe separation distance between the electrical hazard and any part of the lifting equipment (3.1 meters/10 feet for voltages up to 50,000 volts, 4.6 meters/15 feet for 50,000 volts – 200,000 volts, 6.1 meters/20 feet for 200,000 volts – 350,000 volts and 7.6 meters/25 feet for over 350,000 volts);
- The equipment operator should not leave the operating controls while the load is suspended;
- Each personnel involved in the lifting operation should not undertake more than one activity at a time;
- Strict access control should be enforced where lifting operation is being undertaken. The work area, including the swing radius should be demarcated with barrier tape and warning signs erected. No one should work or be under a lifted load. Loads should not be moved directly above people;
- Personnel should have escape routes in case of emergency.

CHAPTER 11

PRESSURE TESTING

Pressure testing is a process of applying stored energy to a facility or equipment (such as pipelines, pressure vessels, boilers, gas cylinders) in order to test its strength, integrity and verify leaks. It involves filling the equipment with a fluid and pressurizing it to a specified test pressure and checking whether or not the pressure is maintained within a specified period. It is a common industrial procedure conducted in petroleum and other process industries.

Pressure testing is a high-risk activity and it is important to carry out all the safety procedures before, during and after the testing. The high pressure can explode either the equipment under test or the test assembly, creating flying fragments and the test fluid projected under immense force with potential for serious injuries to exposed workers and assets damage.

Types Of Pressure Testing

Hydrostatic Testing – This is using water as the test medium and it is usually the preferred type of pressure testing. In hydrostatic testing, water cannot be compressed and hence less stored energy is involved and less hazardous than other test fluids.

Pneumatic testing – this is using air as test medium and it should be done only when hydrostatic testing is not feasible. In addition, for equipment in hydrocarbon service air should only be used when the system has been cleaned to prevent explosive air-hydrocarbon mixture. Coupled with that, the stored energy in pneumatic testing is very high and hence has greater potential for hazard. Unlike hydrostatic testing, pneumatic testing leads to air compression to high pressures and therefore highly risky for workers.

Inert gases (like nitrogen or helium) and steam are also used for medium pressure tests such as tightness and services tests. It is not advised to use oxygen or toxic gases due to the high risk they constitute.

Pressure Testing Hazards

Much of the hazards inherent in pressure testing are associated with uncontrolled release of the stored energy in the test medium. A pressure testing equipment and the system being tested could have joints, valves, gauges, and other components. Uncontrolled release of stored energy from such equipment may result in catastrophic failure with these components flying in varying directions with very high velocity. Such projectiles hitting workers or equipment have potential consequences of major injuries, fatalities and asset damage. In addition, the released fluid hitting personnel at such high velocity has potential for serious bodily harm, and fatality.

In situation where an inert gas is the test medium, release of such in a work location may lead to oxygen deficiency, asphyxiation and death.

Water flooding in an electrically active site may lead to electrocution and consequent fatality.

There are also hazards associated to personnel exposure to chemical additives to the test medium.

The most frequent injuries recorded during pressure testing activities are damage to the eyes, laceration, skin rupture, fractures, damage of internal organs, contusion and concussions.

Precautionary and Control Measures

The following are the common failures that increase the risks inherent in pressure testing:

- Lack of proper pressure testing equipment;
- System over pressurizing;
- Improper test sequence;
- Use of wrong test medium;
- Human error;
- Improper tampering with the system while under pressure;
- Failure to segregate the area in which pressure testing is being performed.

The following should be followed in addressing these common failures:

- Specific procedure must be developed for each pressure testing activity. The development should be done by competent engineers, reviewed and approved by plant manager or his assignee. The procedure will, as a minimum, specify the test medium, temperature of the medium, chemical additives (if required), test manifold arrangement (with drawings), test pressure, relief valve size and set pressure, test sequence, range of pressure gauges and locations, access control requirements, method of disposal of test medium, emergency response procedure, material safety data sheet of chemical additives, hold point for inspection, personal protective equipment and work permit requirements.

- Job hazard analysis must be conducted summarizing the hazards and the controls that must be in place before, during and after the test. This has to be communicated to all personnel involved in the work and adequate resources provided for effective implementation.
- The test manifold should be designed, fabricated and inspected by competent persons. It should be tested to a pressure of at least 20% above the maximum test pressure to be applied. Re-inspection and re-test of the manifold is advisable to be conducted preferably once every 12 months and appropriately tagged with test due dates. It shall not be used past the due calibration date.
- Calibrated relief valves with adequate capacity to prevent over pressurizing should be installed on the test manifold and the system being tested. There should be no isolation valve between the relief devices and the system on which they are installed. Where this is unavoidable, the isolation valves should be car-sealed in open position. The relief outlet should be oriented such it can discharge towards a safe location.
- Reliable and calibrated pressure gauges are required to be installed on the test manifold and the system being tested. The calibration should be done within 30 days of the pressure test and the maximum text pressure within 30 to 80 percent of full range of the gauges. The pressure gauges should be installed in positions that they are easily readable by personnel performing the pressure test.
- The equipment being pressure tested is required to be positively isolated or completely disconnected from the rest of the system or plant.
- All workers that have parts to play in the activity must be trained in their respective roles, hazards of the operation,

approved pressure testing procedure and other hazards control measures.
- Pressure testing should be supervised by competent personnel.
- The material safety data sheet (MSDS) of the test fluid and all chemicals to be used should be on site and the specified precautions followed.
- The test site should be marked, barricaded and warning signs posted at suitable locations.
- Only workers involved in the activity should be allowed into the barricaded area.
- The test must be performed following the specifications and sequence in the approved procedure.
- The system under pressure should be depressurized prior to any adjustment or repair works.
- Adequate lighting should be available throughout the testing activity.
- Safety equipment and supplies should be readily available. Examples are emergency spill kit, fire extinguisher, whip checks and first aid kits.
- On conclusion of the pressure test, the system should be safely de-energized and test medium disposed off in line with the approved procedure.

CHAPTER 12

ELECTRICAL SAFETY

Electricity is a common source of energy in the petroleum industry. However, it has potential for fatalities, serious injuries and extensive property damage if not safely used. During turnaround maintenance activities, people work with or near electrical equipment. This chapter is about the hazards associated with such activities and precautionary measures to significantly reduce the risk of the hazards resulting into incidents.

Electrical Hazards

The major hazards of working with electricity or near electrical equipment and the most potential incidents are as below.

- Electric shock and burns -Shock happens when a body part makes contact with electricity and becomes part of the electrical circuit. The current goes into the body at one point and exists from another. Electrical shock is a reflex response to this passage of electric current through the body. This could cause electrical interference or damage to the heart and cause it to stop or beat erratically. It could also cause burns

to the skin or internal tissues. There are also burns resulting from body contact with overheated electrical equipment or cables. Electrocution is when electric shock leads to fatality.
- Arc flash – Arc flash results when there is a sudden release of electrical energy through the air when a high-voltage gaps exists and the current jumps from one conductor to another or the ground. This gives off huge amount of heat (in the form of thermal radiation), loud explosion and intense, bright light that has potential for burns. Temperatures generated in an arc flash can reach as high as 35,000°F. Hearing loss can also result from the sound of the explosion.
- Fire – resulting from faulty electrical installations or equipment.
- Explosions – The use of electricity can sometimes generate hot surfaces and sparks. In situations where explosive atmospheres exist these can lead to ignition. During turnaround maintenance, explosive atmosphere can be present in several places like paint booths, in sumps, during isolation or de-isolation, in sumps, inside hydrocarbon vessels and many other places.
- Static electricity – igniting flammable vapors or dusts.
- Fall from heights – the reflex reaction from electric shocks or startle reaction or force of explosion from arc flash can cause a worker to fall from heights (for examples, workers on ladders or scaffolds).

Most of the electrical incidents happen due to one of the following three scenarios:

- Unsafe equipment or installations;
- Unsafe work practices;
- Unsafe work environments.

Precautionary and Control Measures

In-order to minimize or eliminate the risk of these incidents, the following measures should be adopted when working on or in close proximity to electrical conductors and equipment:

- When work is to be done on an electrical conductor or equipment, it should be considered energized unless it has been isolated, locked, tagged and verified (for effective isolation). All sources of electrical energy must be identified and shut-off. People will now be cleared from the impacted areas and the equipment or circuitry tested to ensure they cannot be re-energized by any other means. Residual energy (such as, in capacitors) should also be depleted by bleeding, blocking or grounding. It is important to stress that isolation of electrical equipment must be done by a competent person responsible for the plant or section there-of. Locks and tags are to be placed on isolation devices or points by the responsible competent person and the keys kept in a secure location to prevent the equipment or circuit from being re-energized without the knowledge of the person that placed the locks. Maintenance or contractor crews that have activities to perform on them will also place their respective locks and tags on each isolation points.
- If high voltage equipment cannot be effectively isolated before maintenance or service is carried out on it, only personnel that have received adequate training and have the required competence should be allowed to work on them.
- Circuit breakers and fuse boxes should be clearly labeled (and always updated) as to which equipment they are for, so that quick isolation can be done in case of emergency.
- Personnel should never open or close a disconnect switch while facing it. They should stand to the side, faces turned away from the switch, and operate it with a quick, single motion.

- Use of jewelry or articles of clothing that have exposed metallic components should be avoided while working on or in close proximity to energized electrical equipment.
- Access to panels and circuits breakers or fuse boxes should always be kept clear.
- Arc flash warning signs should be affixed on high voltage equipment designating approach boundaries and the required PPEs.
- Only qualified personnel should be allowed to repair or install electrical equipment.
- Personnel who work on live high voltage equipment should follow the requirements of the arc-flash hazard warning signs placed on it by wearing the required PPEs, use insulated tools and other safety related precautions.
- Electrical equipment (including extension cords, lightings, portable tools) brought into hydrocarbon facilities should be explosion proof type. Risk assessment should be conducted on the types of work activities, equipment to be used and the area of the facilities before and measures to control the risk posed put in place. Electrical equipment to be used in potentially explosive atmospheres must be designed and constructed such that the risks of ignition are eliminated or reduced. The equipment must be marked with an 'Ex' to indicate they are suitable for use in potentially explosive atmospheres. The 'Ex' marking is followed by letters and numbers specifying the equipment group, category and characteristics of the explosive atmosphere for which the equipment has been designed. Therefore, pre-mobilization inspection of such equipment should be conducted by competent persons to ascertain they meet the requirements of the specific facility and area.
- Portable cord-and-plug connected equipment, extension cords, power bars and electrical fittings should be inspected for damage or wear before each use. Damaged equipment should be repaired or replaced immediately.

- Extension cords or equipment to be used should be checked to ensure that they are rated for the level of amperage or wattage that they are being used for. Extension cords should not be used in place of permanent installation.
- The correct size of fuse should always be used. Replacement of a fuse with one larger in size has the potential to cause excessive currents in the wiring with consequences of fire.
- Metal ladders should not be used when working on or in the vicinity of live electrical equipment.
- Ground fault circuit interrupters (GFCIs) should be used for all portable electrical tools, temporary wiring and when working in areas that are wet or damp. GFCIs are designed to protect people from electrical shocks. It detects ground faults and quickly interrupts the flow of electric current thereby limiting the duration of a potential electrical shock.
- Exposed receptacle boxes should be made of non-conductive materials.
- When heavy equipment is to work close to overhead power lines, the minimum safe distances should be as follows: up to 33,000 volts, 3 meters (10 feet); 33,000 volts to 275,000 volts, 6 meters (20 feet); above 275,000 volts, 8 meters (27 feet).
- Care should be taken to prevent static discharges in potentially explosive atmospheres. Measures such as earth bonding and the selection of antistatic work clothing and footwear can help to reduce the risk of static discharges.
- Workers that are exposed to electrical hazards should be well trained on their specific crafts, the hazards they are exposed to, precautionary measures and appropriate emergency response procedures. Most importantly, adequate resources must be provided to implement the precautionary and emergency response measures.

CHAPTER 13

PAINTING AND COATING

Paints and coatings can be applied either as solids, liquids or aerosols. They are mainly used during turnaround maintenance for protection of plant and equipment from corrosion.

Painting and Coating Hazards

Paints are chemicals products and whichever ways they are applied have health, fire and explosion hazards. Workers can be exposed to health hazards by inhalation, ingestion or absorption of the paint and coating materials. The hazards stem from the chemicals used in making the solvents, pigments and primers, the common ones being toluene, xylene, ketones, esters, alcohols, lead, chromium, nickel, cadmium, zinc, epoxy resins and isocyanates. The main health effects from exposure to these chemicals include: eye and skin irritation; respiratory tract irritation; dermatitis; dizziness; nausea/vomiting; heavy metal poisoning. Chronic exposure can cause long term health effects such as nerve, kidney or liver damage.

The major safety concern associated with painting and coating is the combustible and flammable vapors, mist and residues that are created. Potential sources of ignition of these include: open flames; cutting and welding activities; heating units; electrical sparks; static electricity; smoking.

Precautionary and Control Measures

The following safety measures should be in place in order to minimize the risks associated with painting and coating operations.

- If possible, the paint type should be substituted with non-flammable and less toxic types.
- When painting or coating activity is being conducted in spray booths or rooms, adequate ventilation should be in place.
- Mechanical ventilation equipment should be inspected regularly to ensure they function properly.
- When spray painting is being carried out in a place other than in a spray booth, it should be conducted at least 6 meters from anything that may block ventilation. Otherwise, the site should be adequately ventilated and all sources of ignition should be removed or isolated.
- Adequate measures should be taken to ensure other workers not involved in the painting activity are informed and that they are not exposed to the paint and coating materials.
- Prior to painting and coating activities, the material safety data sheets should be used to train workers on the associated hazards, and recommended control/recovery measures.
- Correct storage, transportation, use and disposal requirements (as specified in the material safety data sheet) should be followed.
- Appropriate personal protective equipment (including respiratory protective equipment) should be worn by all personnel carrying out painting and coating or who are exposed to the materials (as specified in the materials safety data sheet).

- Functional eye wash and safety shower facilities should be available within the vicinity of painting and coating sites and materials storage areas.
- The painting and coating equipment should be properly grounded.
- There should be adequate safety signs (especially, restricting ignition sources within the painting zone).
- Flammable paint and coating materials should not be applied on hot surfaces.
- Painting and coating jobs should not be carried out up wind of an ignition source; unless there are adequate measures to ensure that flammable vapors do not drift to the ignition source.
- Spray painting booths should be regularly cleaned of residues with non-sparking tools.

CHAPTER 14

HAND AND POWER TOOLS

There are various types of hand and power tools used during turnaround maintenance activities. The following are some of the general inherent hazards:

- Body and eye injuries due to projectiles from tools chips or materials being worked on;
- Cuts, punctures, abrasions and fractures;
- Noise;
- Vibration;
- Electrical;
- Moving parts.

Generally, there are some basic safety rules that can help to control hazards associated with the different types of hand and power tools:

- All hand and power tools should be kept in good condition and regularly maintained;
- The right tool should be selected and used for the task it is made for;

- Tools should be inspected for damage before use and damaged ones removed promptly from site. They should be re-inspected after use before storage;
- Tools are required to be operated and used according to manufacturers' specifications;
- The required type of personal protective equipment should be worn while using hand and power tools.

Specific Precautionary and Control Measures

Hand Tools – Hand tools are tools powered manually and they include screw drivers, pliers, hammers, chisels, saws, spanners, wrenches etc. These are so commonly used that people do not think that they pose hazards. However, serious incidents can occur if tool-related hazards are not effectively controlled. The additional safety precautions necessary for the use of hand tools are as follows:

- Workers should be trained in the safe use, maintenance and storage of hand tools;
- Cutting tools must be sharp; dull tools can cause more hazards than sharp ones;
- When using tools like saw blades and knives, they should be directed away from aisle areas and other employees working in close proximity;
- When used for electrical work, handles of tools should be insulated;
- When working in locations where flammable gases exist or are likely to exist, non-sparking tools made of non-ferrous materials should be used;
- The use of 'cheater bar' extensions in spanners, and wrenches should be discouraged;
- Handles of screw drivers should not be subjected to hammer blows;
- The head of hammers should be properly secured by wedges;

- Homemade tools should not used at the work sites;
- Appropriate personal protective equipment such as safety goggles and hand gloves should be worn while using hand tools;
- Proper housekeeping should be maintained in the workplace to prevent accidental slips with or around dangerous hand tools.

Power Tools – There are five basic types of power tools, differentiated by their power source and they are: electric, pneumatic, liquid fuel, hydraulic and powder-actuated. All power tools should be fitted with guards and safety switches.

To control hazards associated with the use of power tools, workers should follow the following general precautions:

- Tools should never be carried by the cord or hose;
- Cords or hoses should never be yanked in order to disconnect the tool from the energy source;
- Cords and hoses should be kept away from heat, oil and sharp edges;
- All tools should be switched OFF before connecting them to a power source;
- Isolation and effective lockout from energy source should be done before carrying out any maintenance work tasks or making adjustments on a power tool;
- Power tools should be properly grounded or double-insulated and should be tested for effective grounding with a continuity tester or a Ground Fault Circuit Interrupter (GCFI) before use;
- The on/off switches should not be bypassed to operate the tools by connecting and disconnecting the power cords;
- Electrical equipment should not be used in wet conditions or damp locations unless it is connected to a GCFI;

- Work pieces should be secured with clamps or vise, freeing both hands to operate a tool;
- Tools should be maintained with care, keeping them sharp and clean for best performance;
- When lubricating and changing accessories, the manufacturers' manual should be followed;
- When operating power tools, good footing and balance should be maintained;
- Loose clothing, ties, jewelry and long hair can become caught in moving parts and hence should be prohibited while working with or in close proximity to power tools;
- All damaged power tools should be removed from service and tagged-Do Not Use;
- The exposed moving parts of power tools (such as belts, gears, shafts, pulleys, sprockets, spindles, drums, flywheels, chains, or other reciprocating, rotating, or moving parts) should be safe guarded and safety guards should never be removed when a tool is being used;
- Hand-held power tools should be equipped with a constant-pressure switch or control that shuts off the power when the pressure is released.

Electric Tools–the most serious risks associated with the use of electric tools are burns and shocks. The following additional general practices should be followed when using electric tools:

- To protect users from shock and burns, electric tools should have three-wire cord with a ground and be plugged into a ground receptacle, be double insulated or be powered by a low-voltage isolation transformer. Otherwise, double-insulated tools are available that provide protection against electrical shock;
- Electric tools should be operated within their design limitations;

- Hand gloves and appropriate safety shoes should be worn when operating electric tools;
- Electric tools should be stored in a dry place when not in use;
- Work areas should be kept well-lit when operating electric tools;
- Cords from electric tools should be kept such that they do not present a tripping hazard.

Pneumatic Tools– the following additional safety precautions should be followed when using pneumatic tools:

- They should be inspected to ensure that the tools are fastened securely to the air hose to prevent them from disconnecting;
- A positive locking device or short wire attaching the air hose to the tool should be used and serve as an added safe guard;
- For a hose that is more than half inch in diameter, a safety excess flow valve should be installed at the source of the air supply in order to reduce pressure in case of hose failure;
- Hoses of pneumatic tools should be kept such that they do not present a tripping hazard;
- When using pneumatic tools, a safety clip should be installed to prevent attachments (such as chisels on a chipping hammer) from being ejected during tool operation;
- Pneumatic tools that shoot fasteners and are operated at pressures more than 100 pounds per square inch (6,890 kPa) should be equipped with a special device to keep fasteners from being ejected, unless the muzzle is pressed against the work surface;
- Eye, ear, face, head, hand and foot protections should be worn by personnel working with pneumatic tools;
- Compressed air guns should never be pointed at people;
- Screens should be set up to protect nearby workers from being struck by flying fragments around chippers, riveting guns, staplers or air drills;

- To reduce the extent and duration of continuous exposure to vibration, job rotations or more frequent breaks should be practiced by workers operating jack hammers.

Liquid Fuel Tools – the most serious hazards associated with the use of fuel-powered tools are from the flammable fuel vapors (with the attendant risk of fire/explosion or asphyxiation by exhaust fumes). Hence, adequate care should be taken in the handling, transportation and storage of the fuel. In addition:

- The engine powering the tool should be shut down and allowed to cool before refueling;
- There should be adequate ventilation if liquid fuel tool is being used in a closed area to avoid carbon monoxide inhalation;
- Liquid fuel tools should not be positioned upwind of work location in an open area;
- Adequate fire extinguishers should be available at the work location.

Hydraulic Power Tools – The following safety requirements should be in place:

- The fluid used in hydraulic power tools should be fire resistant;
- The specified safe operating pressure for hoses, valves, pipes, filters and other fittings should not be exceeded;
- Hydraulic jacks should have stop indicators and the stop limit should not be exceeded;
- The load limit for jacks should be written on it and should never be exceeded;
- Jacks should never be used to support a lifted load;
- Jacks should be lubricated and maintained regularly. They should be inspected at least once every six months and when subjected to abnormal loads or sent out of the shop for special use, should be inspected before and after.

- **Powder-Actuated Tools**– The safety precautions that should be in place when using powder-actuated tools are as follows:
- Only trained employees should operate powder-actuated tools;
- Users of powder-actuated tools should wear ear, eye and face protection;
- The tool should not be used in an explosive or flammable atmosphere;
- Users should inspect the tool before using it to determine that it is clean, that all moving parts operate freely and that the barrel is free from obstructions and has the proper shield, guard and attachments as specified by the manufacturer;
- The tool should not be loaded unless it is to be used immediately;
- Loaded tool should not be left unattended;
- The tool should never be pointed at people;
- Fasteners should neither be fired at materials that are so soft that the fasteners could pass through to the other side nor at materials that are so hard or brittle that the fasteners might ricochet or the material splatter;
- Alignment guide should always be used when shooting fasteners into existing holes;
- When using a high velocity tool, the fasteners should not be driven more than 3 inches (7.62 centimeters) from an unsupported edge or corner of material such as brick or concrete. And fasteners should not be placed in steel any closer than half an inch (1.27 centimeters) from an unsupported corner edge unless a special guard, fixture or jig is used.

CHAPTER 15

CUTTING, WELDING AND BRAZING

The main hazards associated with cutting, welding and brazing are:

- Electric shock;
- Overexposure to fumes and gases;
- Arc radiation;
- Fire, explosion and burns.

Precautionary and Control Measures

The following are the safety requirements to ensure that these hazards are controlled:

- Welding equipment should be undergoing regular preventive maintenance checks as recommended by the manufacturers;
- Equipment should be inspected before use on a daily basis and defective ones should be put out of service until the defect is rectified;
- The following components of the equipment should be checked: all connections tight; output terminals insulated;

electrode holder and welding cables well insulated; and settings correct for the job to begin;
- For engine-driven machine, it should be checked to be running properly, all hoses tight, fuel cap tight, and no fuel or oil leaks;
- Welders should wear proper PPEs that should include flame-resistant clothing, aprons, leggings, leather sleeves/shoulder capes, leather gloves, safety shoes, hard hats and welder's face shield;
- Electric welding machines should be properly grounded;
- Areas where welding activities are ongoing should be properly ventilated. When carried out in workshops, mechanical ventilation should be used;
- Welding, cutting and brazing should not be carried out in oxygen enriched atmospheres (concentration greater than 23.5 %);
- There should be good housekeeping around areas where welding, cutting and brazing are being conducted. Combustible materials should be removed and those that cannot be removed should be covered with fire blankets;
- Sewers within 25 meters of the welding location should also be covered;
- Portable fire extinguisher should always be available where welding, cutting and brazing operation is being conducted;
- A trained fire watch should man the extinguisher, watch out to extinguish any incipient fire while the operation is ongoing and at least 30 minutes after;
- When flame-cutting or brazing, lighted torches should not be left unattended;
- Oxygen/acetylene cutting sets should be checked before use for leaks (for example, with soap solution);
- Oxygen/acetylene cutting sets gauges, regulators, hose, cylinders should be checked for damage and the cylinder should have valid hydro-test date;

- Flashback arrestors should be installed at oxygen and acetylene cylinder regulators;
- Check valves should be installed at the torch end of the hoses;
- Pressure regulators should have functional gauges;
- Oxygen and acetylene gas regulators should be turned off when not in use;
- Gas hoses should be protected from damage during operations;
- Oxygen/acetylene torches should be lit with strikers only;
- Acetylene cylinders should have a handle or valve wrench at all times;
- When attempting to stop gas leaks, the cylinder valve should be closed first;
- Oil or grease should never be used as lubricant on cylinder valves or attachments;
- Gas cylinders should be protected from direct sunlight, flame or any sources of heat;
- Cylinder protective caps should be in place when the cylinders are not in use;
- Cylinders should be transported in upright position, in trolleys or material baskets or cylinder racks or other proper cages;
- Cylinders should never be moved with slings or rope;
- While moving cylinders, they should never be knocked on each other;
- In storage, cylinders should be in upright position in cylinder racks;
- Oxygen and acetylene cylinders should not be stored together. There should be at least 20 feet (6.1 meters) separation distance between them;
- Where the above separation distance is not possible, there should be 5 feet (1.5 meter) non-combustible wall between oxygen and acetylene cylinders;
- Stored cylinders should be clearly labeled (empty and full cylinders properly identified);

- Cylinders should not be stored under direct sunlight or on bare ground;
- Gas cylinders should be stored in well-ventilated locations;
- Gas cylinders should never be stored at temperatures above 54 degrees Celsius (130 degrees Fahrenheit).

CHAPTER 16

IONIZING RADIATION

Ionizing radiation is used to investigate the integrity of structures or components through radiographic images, generally referred to as non-destructive radiography or industrial radiography.

Hazards of Ionizing Radiation

Overexposure to ionizing radiation has serious consequences like damage of body cells, leukemia or solid tumor (cancer). In pregnant women, the exposure of the embryo or fetus to ionizing radiation could increase the risk of leukemia in infants, mental retardation, and congenital malformations.

Precautionary and Control Measures

The following safety precautions (as a minimum) should be in place to ensure that radiation risk is reduced to as low as reasonably achievable (ALARA):

- Procurement, handling, use, storage and disposal of ionizing radiation equipment and consumables is a legal issue and should be conducted in line with local regulation;
- Every ionizing radiation activity should be conducted under the supervision a radiation safety officer (RSO), who should be a certified industrial radiographer (the level of which should be in accordance with the local regulation);
- All industrial ionizing radiation equipment should be registered with the local regulatory authority and the evidence of such registrations should always be available on site;
- No personnel should be allowed to operate ionizing radiation equipment without proper training, licensing and authorization (in line with local regulation);
- Industrial radiographers should be assigned with and wear photon-sensitive passive dosimeters as well as instantaneous reading electronic alarm dosimeters;
- Each passive personal dosimeter should be worn by only the individual to whom it has been assigned in order to keep the personnel monitoring data;
- The above data should be retained as a permanent record and be made readily available for review regularly to ensure they do not exceed allowable threshold value;
- As much as possible, radiography should be performed at periods when human traffic is least at the site (most likely in the night);
- Prior to the commencement of radiography, a controlled area should be established and access into such area should be controlled with barriers, warning signs (visual and audio) and safety instructions strategically posted external to the area;
- The controlled area should also be monitored by designated individuals to enforce the access restriction;
- Calibrated and functional survey meters should be used to ensure that radiation dose rate at the boundary of the controlled area is within safe limits;

- Temporary radiation control station should be positioned outside the controlled area for the initiation, generation or termination of ionizing radiation and for image acquisition and assessment;
- Personnel who are working in close proximity to the controlled area should be told about the activity, the associated hazards and controlled measures in place;
- Transportation and storage of radioactive sources should be conducted as approved by local regulations in order to prevent undue exposure to the public, theft or loss and accurate inventory should be kept at all times;
- Conspicuous warning signs should be displayed on vehicles transporting radioactive sources, and at the storage locations;
- Disposal of radioactive waste should also be done in accordance with local regulations;
- Accidents involving radioactive sources or equipment (including theft) should be reported immediately to regulatory authorities and steps taken to give medical attention to any exposed people;
- Adequate investigation of such accidents should be conducted to prevent reoccurrence;
- Repair and maintenance of radiation equipment should be carried out only by competent and licensed personnel who should ensure effective isolation is in place before carrying out the job.

CHAPTER 17

WORK AT HEIGHTS

Work at heights remains one of the leading causes of fatalities and serious injury. 'Work at height' means work in any place where, without adequate precautions in place, a person could fall a distance capable of causing personal injury or even death.

Precautionary and Control Measures

Work at heights should be properly planned, supervised and carried out by competent people with the skills, knowledge and experience to do the job. The risk should first be assessed and the following factors weighed: the height of the work surface; the duration and frequency; and the condition of the surface being worked on. Before working at height, it is important to work these simple steps:

- Avoid work at height where it is reasonably practicable to do so;
- Where work at height cannot be easily avoided, prevent falls using either an existing place of work that is already safe or the right type of equipment;

- Minimize the distance and consequences of a fall, by using the right type of equipment where the risk cannot be eliminated.

The following are safe methods of working at heights:

Personal Fall Arrest Systems

A personal fall arrest system is one option of protecting personnel if the working height is greater than 6 feet (1.8 meters) from grade. It is comprised of three (3) key components – anchorage/anchorage connector; body wear; and connecting device.

Anchorage/Anchorage Connector – anchorage is commonly referred to as a tie-off point (examples: lifeline, I-beam, rebar, scaffolding) and anchorage connector is the connecting device to the anchorage (examples: cross-arm strap, beam anchor, D-bolt, hook anchor). An anchorage should be capable of supporting 5,000 pounds of force per worker. It should be high enough for a worker to avoid contact with a lower level should a fall occur. And the anchorage connector should be positioned to avoid a "swing fall".

Body wear is the personal protective equipment worn by the worker. In the petroleum industry, the only form of body wear acceptable by many companies for fall arrest is the full-body harness.

Connecting Device – this is the critical link which joins the body wear to the anchorage/anchorage connector. Examples are shock-absorbing lanyard, fall limiter, self-retracting lifeline, and rope grab. The potential fall distance should be calculated in order to determine the type of connecting device to be used. For distances under 18.5 feet (5.6 meters), self-retracting lifeline/fall limiter should be used. Shock-absorbing lanyard or self-retracting lifeline/fall limiter should be used when the height is over 18.5 feet (5.6 meters).

When using personal fall arrest systems, the following measures are important:

- Users should inspect the device prior to each use and defective components discarded;
- The components of the device should be protected against cuts or abrasions and should not be used to hoist materials;
- A device should not be used after stopping a fall as the components could have weakened.

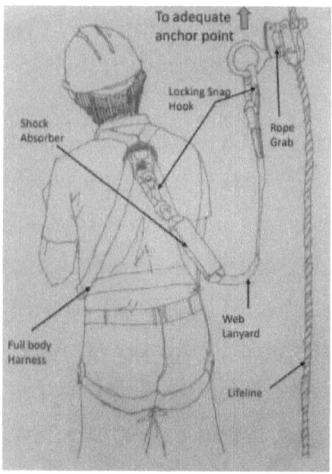

Figure 15: Drawing showing different parts of full-body harness

Figure 16: Work at height with full body harness

Scaffolds

Scaffold is a temporary work platform used to support personnel and materials when working at height. Scaffolds should be properly designed, erected and inspected by competent and certified personnel.

There are various types of scaffolds and some of them are: supported, suspended, tower and mobile scaffolds. The most commonly used form is the supported scaffold that is built from the base upwards. The standard components are as follows:

- Sole boards – these support the scaffold post on soft surfaces to ensure it does not sink and loose balance;
- Base plate- these are installed below the posts, sometimes with screw jacks, if there is a need to even out the levels;

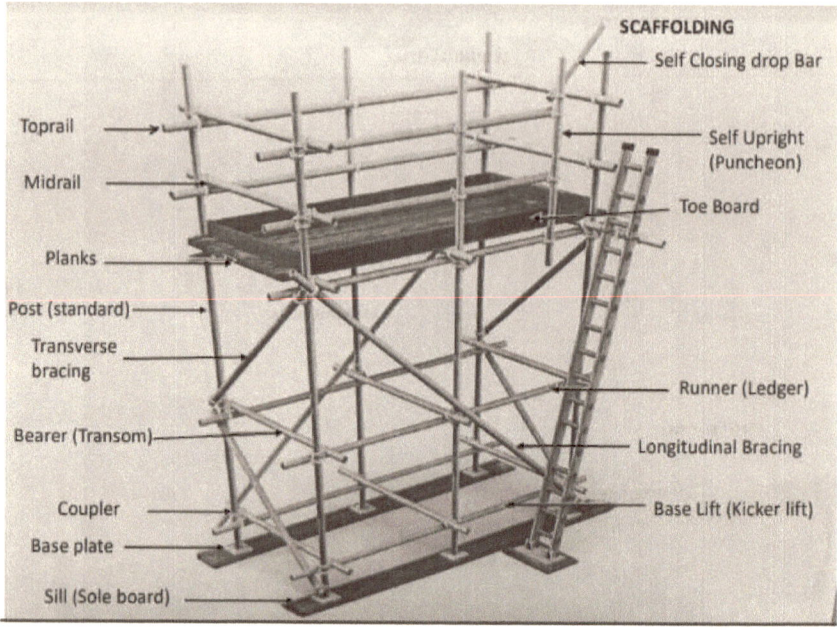

Figure 17: Scaffold showing the various parts

- Post – these are the vertical standards that support the scaffold structure;
- Bearers- these the horizontal supports that run across the width of the scaffold and on which the working platform rests;
- Runners – these are the horizontal supports of the working platform that run along the length of the scaffold. The bearers rest on top of the runners;
- Planks (boards) – these form the working platform;
- Toe board – these rests on the working platform at all open ends to prevent materials from falling on top of people below;
- Longitudinal bracing & transverse bracings - these keep the scaffold structure stable;
- Top rail – this is the horizontal guard rail installed at the height of between 38 and 45 inches (0.95 – 1.15 meters) above the working platform;

- Mid rail - this is the horizontal guard rail installed midway between the top rail and the working platform;
- Access and egress Ladder – this is installed for safe access and egress from the platform.

Figure 18: Scaffold with yellow tag

Scaffolds should be erected by personnel trained and competent in erecting scaffolds. The erection should be supervised by experienced and competent scaffold supervisors. During and after the erection of a scaffold, it should be appropriately tagged by a certified and authorized scaffold inspector. Before using a scaffold, it is important to check the components are complete and that it is tagged. A scaffold tag gives its safety status. Scaffold tags come in three colors: green, yellow and red. A green tag indicates that the scaffold is safe for use within the validity period indicated on it. The tag will also specify the maximum allowable load. A scaffold that is past its validity date should not be used until it has been re-inspected. A yellow scaff-tag indicates it is not safe to use without a fall arrest system within the specified validity period. A scaffold without a tag or with a red scaff-tag should not be used.

The simple tips for the safe use of a scaffold are as follows.

- It should be inspected each day (using an approved checklist) before use to ensure it has the correct tag and that it has not been tampered with.
- No part of the scaffold should be removed, otherwise, it must be re-inspected by a certified person.
- It should not be loaded beyond the capacity specified on the tag.
- It should be re-inspected after every rain or heavy wind.
- Personnel should not climb the braces but use the ladders.
- Unsafe areas underneath the scaffold should be roped-off to prevent passers by being hit by falling objects.
- If it is not practicable to rope off the areas underneath, wire mesh should be provided around the work area.

Elevating Work Platforms

An elevating work platform (EWP) is a mechanical device that provides a safe platform when working at heights. Scissor lifts, cherry pickers, boom lifts and travel towers are all types of EWPs. They can be battery powered or make use of internal combustion engines.

The safety tips required to work with EWP are as follows.

- Only qualified and authorized personnel should operate EWPs.
- Pre-startup check should be performed on EWP at least once every day and documented.
- EWP should not be operated closer than 10 feet (3 meters) to power lines with voltage up to 50KV, 20 feet (6 meters) for

voltages between 50 KV and 250KV, and 25 feet (7.5 meters) for voltages above 250 KV.
- Personal fall arrest devices should be worn at all times while inside EWP.
- Personnel should never enter or leave EWP when it is elevated and should remain within the confines of the work platform at all times.
- Personnel should not stand on the handrails of the platform.
- EWP should not be used as a crane or hoist and materials should not be slung or attached to the side of an EWP, unless the attachment has been specifically designed, rated, tested, marked and approved by the manufacturer of the EWP.
- EWP should not be operated on uncompacted ground.
- Outriggers should always be used in EWP that have them.
- Emergency descent switches should be identified and tested prior to operation.
- Ground controls and descent valves should be known by all personnel.

Ladders

When the work at height is short time in nature and does not require much physical exertion, it may be advisable to use a ladder, provided the following controls are in place.

- Employees should complete inspection of the ladder and ensure that it is in good condition before each use.
- If it is possible, use only ladders that are Underwriter's Laboratory approved (will have UL seal).
- Ladders should not be painted as paints can hide defects.
- Step ladders should be fully opened and locked before climbing.
- Ladders should be placed on flat, secure, hard and non-movable surfaces.

- It should not be placed at the back of a closed door.
- The base should be positioned one foot away for every four feet of height to where it rests (1:4 ratio) and be properly secured at the top.
- The rails should extend at least three feet above top landing.
- If work is to be done at a height more than 6 feet (1.8 meters) above the ground or working surface, scaffold or EWP should be used instead.
- Before climbing, shoes should be checked to ensure they are free of grease, mud or anything that could cause slippage.
- Users should face the ladder when ascending or descending and ensure at least three points of contact with hands and foot at every moment.
- Tools should be carried in pockets or bags attached to a belt or raised and lowered by rope.
- Workers should not climb higher than the third rung from the top.
- While working, the user should face the ladder and maintain three points of contact.
- User should not overreach, but always keep his/her torso between the ladder rails.
- It should not be used outdoor on windy days.
- Metallic ladders should not be used around electrical conductors or equipment.
- Only one person is allowed on top of a ladder at a time.

Fall Protection Plan

Each work activity that is required to be performed at height should be identified and fall protection plan developed. The fall protection plan should include: identification of fall hazards of the activity; appropriate method of access and fall protection equipment to utilize; selection of anchorage point (if personal fall arrest equipment is to be used); methods of securing lanyards and lifelines; method of protection of other workers from falling objects; and emergency response procedure.

Inspection of these equipment should be carried out to ensure that they are in good conditions and safe to use. It is recommended to have a color coding for ease of identification of inspected equipment. The re-inspection period and criteria should be specified in the fall protection.

Personnel that are required to work at heights and their supervisors shall be trained. The content of the training should include: the identified hazards of the activities; use of the required fall protection equipment; inspection, storage and maintenance of the equipment; and emergency response procedures.

The fall protection plan should include action items, responsible parties and timelines.

CHAPTER 18

WORKING IN EXTREME HOT TEMPERATURES

Workers may sometimes be exposed to extreme temperatures or work in hot environments that will put them at risk of heat stress. Exposure to heat stress may have the consequences of occupational illnesses and injuries. Examples of such are heat stroke, heat exhaustion, heat cramps, or heat rashes. In addition, such exposure can increase the risk of other injuries as a result of either dizziness or fogged-up safety glasses or sweaty palms. Other factors that increase the risk of heat stress are overweight, heart disease, high blood pressure, being over 65 years of age and taking certain medications.

Precautionary and Control Measures

Heat stress precautionary and control measures include engineering controls, administrative (work practice) controls and personal protective equipment (PPE).

In engineering control, the workplace is re-designed in a way that reduces exposure to heat. This could take the following forms:

- Increase in air velocity or cooling of the work area- for examples, installation of air-conditioners or air-blowers depending on the configuration of the work space;
- Shielding from direct sunlight – for example erection of shades that minimize exposure to direct sunlight when work is on-going;
- Provision of cool and shaded rest shelters – these should be located such that no worker will have to travel more than 100 meters to a rest shelter.

Administrative controls include training and changes to tasks or schedules to reduce exposure to heat stress. This should include the following:

- Training workers on the risk of heat stress, control measures, symptoms of heat-related illnesses and first aid/recovery measures;
- Regulating the time in the heat and time in a cool rest shelter, depending on the ambient temperature, relative humidity and type of work being done;
- Increasing the number of workers per task;
- Provision of adequate quantity of cool drinking water and encouraging workers to re-hydrate often;
- Use of buddy system so that workers can look out for the welfare of each other when exposed to heat stress;
- Rescheduling work to cooler periods of day or night;
- Using work tools that minimize manual strain;
- Establishing a heat acclimatization plan;
- Establishing a heat stress level communication method that alerts workers of danger category and precautionary measures.

Training

Workers and supervisors should be trained on heat stress management that includes the following:

- Hazards, signs and symptoms of heat-related illnesses and appropriate first aid for each;
- Causes of heat-related illnesses and methods of reducing the risk;
- Using weather reports (temperature and relative humidity) to determine heat stress level and determining appropriate precautionary measures;
- Types, uses and care of heat-protective clothing and equipment;
- Meaning and use of acclimatization;
- Effects of clothes, age, drugs, alcohol, obesity on tolerance to heat stress;
- Emergency response procedures for heat-related illnesses.

Acclimatization

New workers, those just returning from vacation and those re-assigned from different work areas where they are not exposed to heat stress should be progressively exposed to hot conditions over a period of 7 to 14 days. During this period, they should be closely supervised and those not physically fit may be given more time to acclimatize.

Hydration

Adequate supply of cool portable water of temperature, <15°C (59°F), should be made available near the work area and in rest shelters. Workers should be provided with individual (or disposal) drinking cups and be encouraged to hydrate often. It is not encouraged to take alcohol, drinks with caffeine or sugar. Workers should be taught not to wait to get thirsty before drinking water but abide by recommended quantities as applicable to ambient temperature and relative humidity. On the average, workers should drink between 1 to 1.5 liters of water every hour, taken in small quantities at a time.

Rest Breaks

Personnel working in hot temperatures should be encouraged to take appropriate rest breaks in provided shelters and hydrate. Work periods should be shortened and rest periods increased, as per the following indices:

- Increase in temperature, humidity and manual exertion of the assigned task;
- With less air movement;
- When protective clothes or equipment is worn.

The heat stress danger chart below can be used to regulate the work and rest periods. The rest period should then be increased when work being done requires manual exertion or there is less air movement or heavy protective clothes are worn.

Other Administrative Controls

It is also very important that from heat stress danger categories "Extreme Caution" to "Extreme Danger" (refer to chart below), buddy system is put in place. No worker should be allowed to work alone. In addition, in categories "Danger" and "Extreme Danger" there should be no work under direct sunlight. Mechanical ventilation or air-conditioning equipment will be required to be introduced.

Ambient Temperature	10%	20%	30%	40%	50%	60%	70%	80%	90%
>50									
50									
49									
48									
47									
46									
45									
44									
43									
42									
41									
40									
39									
38									
37									
36									
35									
34									
33									
32									
31									
30									
29									
28									
27									
26									

Color Code	Danger Category	Work : Rest Periods (Minutes)
	Extreme Danger	20:10
	Danger	30:10
	Extreme Caution	50:10
	Caution	Normal

Figure 19: Heat Stress Danger Chart

Personal Protective Equipment (PPE)

There are wearable PPE that protects workers from effects of exposures to high temperature. They are auxiliary cooling systems or personal cooling systems. Some examples are water-cooled garments, air-cooled garments, cooling vests, and wetted overgarments. These have the effect of increasing the rate of heat removal from the body and thereby decreasing the potential for heat-related illnesses.

Emergency Response for Heat-Related Illnesses

Heat Stroke

Heat stroke has the potential for permanent disability or death if the victim does not receive emergency treatment. It happens when the body is not able to control its temperature, the sweating mechanism stops to work, the body is not able to cool down and the temperature rises very rapidly to about 106°F or higher within 10 to 15 minutes. It is the most serious heat-related illness.

Symptoms of heat stroke include the following:

- Hot, dry skin and no sweating;
- Confusion, weakness and slurred speech;
- Strong and rapid pulse;
- Seizures
- Weakness, nausea and loss of consciousness (coma).

When heat stroke occurs, the following steps should be taken to prevent fatality:

- Call for emergency medical services and stay with the victim until their arrival;
- Relocate the victim to cool shaded area and remove the clothes;
- Using cold water, ice pack or wet cloths, cool the worker quickly;
- Increase ventilation by either mechanical means or hand fan around the victim;

- Apply ice or wet cloths on the head, neck, armpits and groin.

Heat Exhaustion

Heat exhaustion occurs when there is an excessive loss of water and salt, usually through excessive sweating.

Symptoms of heat exhaustion include the following:

- Headache;
- Rapid pulse;
- Nausea;
- Dizziness;
- Weakness;
- Irritability;
- Pale skin;
- Thirst;
- Heavy sweating;
- High body temperature;
- Low urine output.

When a worker has heat exhaustion the following should be done:

- Call medical emergency services and stay with the victim until their arrival;
- Relocate the victim to a shaded area and give cool drinking water;
- Remove outer clothing, shoes and socks;
- Apply ice pack or wet cloths on the head, face and neck;
- Increase ventilation and encourage the victim to hydrate frequently.

Rhabdomyolysis

Rhabdomyolysis (rhabdo for short) refers to a medical condition that is associated with exposure to heat stress and prolonged physical exertion. Rhabdo results in the quick breakdown, rupture, and death of muscle tissues. When this happens, electrolytes and proteins are released into the bloodstream. This has the potential for irregular heart rhythms, seizures, and damage to the kidneys.

Symptoms of rhabdo include the following:

- Severe muscle aches and cramps;
- General weakness;
- Dark brown urine;
- Confusion;
- Fever;
- Difficulty exercising arms or legs.

When a worker has rhabdo the following should be done:

- Stop work, move to a shaded area and drink more water
- Call medical emergency services and request to be checked for rhabdo.

Heat Syncope

Heat syncope refers to fainting or dizziness that happens when standing for too long in a hot environment or suddenly standing up after prolonged sitting or lying.

Symptoms of heat syncope include the following:

- Dizziness;
- Short duration fainting;
- Light- headedness.

When a worker has heat syncope the following should be done:

- Get the victim to sit or lie down in a cool shaded place;
- Give drinking water, to be taken slowly.

Heat Cramps

Heat cramps occur when a person sweats a lot while working in a hot environment. This depletes the body's salt and moisture levels thereby resulting in painful cramps. This may also be a symptom of heat exhaustion.

Some of the symptoms include muscle pain, cramps, or spasms in the abdomen, arms, or legs.

When a worker has heat cramps the following should be done:

- Relocate the victim to a rest shelter and give cool drinking water;
- If the cramps do not stop within 1 hour, call for medical assistance;
- If the victim, in addition, has heart problems, request for medical help as soon as the symptoms manifest.

Heat Rash

Heat rash refers to skin irritation caused by excessive sweating in hot and humid environment.

Symptoms of heat rash include tiny blisters or clusters of pimples on the neck, upper chest, under the breasts and groin.

Workers experiencing heat rash should do the following:

- Apply powder in the affected areas and keep dry;
- If possible, be moved to work in a cooler and less humid area.

CHAPTER 19
PRE-STARTUP SAFETY REVIEW

In the petroleum industry, major accidents have occurred during startups and shutdowns. In a bid to minimize this happening, pre-startup safety review (PSSR) should be carried out prior to introduction of feedstock or energy, after turnaround and maintenance activities. The reviews are made to avoid work place incidents and consequential loss of lives and equipment damage.

The PSSR is an important part of process safety management and is a final step to ensure safety before bringing back an equipment or plant to service after turnaround. It is designed to ensure that all critical areas have been adequately assessed and addressed such that operation can start smoothly and safely. During the review, every part of the facility should be checked to be in a safe position and actions items arising from management of change process have been addressed. It is worthy of note that PSSR, if not properly conducted can result into major disasters. For an example, in 2005 an explosion occurred in BP Texas City Refinery leading to 15 fatalities and 180 injuries and major asset damage. The investigation report concluded that some of the root causes of the incident were defective management of change process (that permitted the placement of contractor trailers very close to a process equipment), inadequate supervision of start-up operations by competent personnel, poorly trained operators, inadequate communications and the use of outdated and ineffective work procedures. Some of the technical failures were blowdown drum that was of insufficient size, inoperative alarms and level sensors in the ISOM process unit and general lack of preventive maintenance on safety critical systems.

In the United States of America, the Occupational Safety and Health Administration (OSHA) has PSSR regulations. In OSHA 29 CFR (Code of Federal Regulations) 1910.119 these regulations specifically state:

1. The employer shall perform a pre-startup safety review for new facilities and for modified facilities when the modification is significant enough to require a change in the process safety information.
2. The pre-startup safety review shall confirm that prior to the introduction of highly hazardous chemicals to a process:

    i. Construction and equipment is in accordance with design specifications;
    ii. Safety, operating, maintenance and emergency procedures are in place and are adequate;
    iii. For new facilities, a process hazard analysis has been performed and recommendations have been resolved or implemented before startup and modified facilities meet the requirements contained in management of change;
    iv. Training of each employee involved in operating a process has been completed.

The review should be conducted by a team of experienced personnel from various disciplines (for examples, engineering, operations, maintenance, safety, inspections) and they should check that:

- All equipment (vessels, tanks, piping, rotating equipment, relief devices, fire protection systems) are mechanically complete in line with design specifications;
- Explosion proof rated electrical equipment have been inspected and still meet the classification of their locations;
- Equipment grounding is in place and continuity tests conducted to ensure effectiveness;
- Other electrical verifications conducted;
- Instrument verifications conducted;
- All critical safety systems are identified, calibrated and tested;
- Process hazard analysis (where applicable) was completed and recommendations addressed;
- Construction/maintenance wastes have been cleared and good housekeeping in place;
- Temporary equipment, offices and storage facilities have been removed from process areas
- Emergency response equipment are in place and adequate;
- Appropriate safety signs are in place;
- Actions arising from MOC (if applicable) have been addressed;
- All required process safety information is available and updated;
- Operating procedures (including startup) are in place, updated and communicated to relevant personnel;
- Training of personnel on new or changed processes/procedures have been completed;
- Plant/equipment isolation is still in place;
- Plant/equipment has been cleaned or purged.

After the review, recommendations to address identified deficiencies should be categorized into pre-startup and post-startup actions. Pre-startup actions must be addressed before the equipment is put into operation. Post-startup actions can be done after operation is started, implying that they are not critical for safe startup. Before startup, personnel with adequate competence and authority (depending on the complexity of equipment or plant) should review PSSR report, actions taken and give approval for startup to proceed.

CHAPTER 20

STARTUP, COMPLETION AND REVIEW

Startup phase of a petroleum process facility after turnaround is the period when the equipment, facility or unit is taken from idle state to normal operations. It presents the highest risk in operation as process safety incidents are much more likely to happen during that phase than normal operations, because of non-routine procedures that are executed during startup.

To minimize the risk of these types of incidents, petroleum process facilities should have updated startup procedures, provide appropriate training to personnel and put in place effective communication. As part of planning for the turnaround maintenance, shutdown and startup procedures should be reviewed, updated (where necessary) and approved. The procedure should have sufficient details and possibly contain checklists and diagrams. Upon approval from the facility manager for startup to commence, the implementation of the procedure should start and operations personnel are required to be informed of their respective roles and responsibilities in the startup. Control room operators on duty must be experienced in the types of control systems they are operating at the phase and be closely supervised. New workers should be given safety orientation on startup and emergency safety procedures. It is important to specify that only necessary maintenance workforce should be in the facility such as mechanics, electricians, instruments technicians, crane operators and a few contractor laborers.

The procedure and checklist should be followed without bypassing critical safety devices during trouble shooting operations until the facility, equipment or unit reaches its normal parameters.

Post startup items arising from pre-start safety review should now be resolved. These are issues like installation of insulation, removal of scaffolds, housekeeping and demobilization of all equipment brought into the facility for the purpose of the turnaround. The safety team should still be very vigilant as serious incidents still do occur at this stage. Materials, contracts and other resources reconciliation are carried out to complete the turnaround activities.

On completion, a meeting should be held to conduct thorough review of the turnaround maintenance activities. This review should be based on the plans and key performance indicators set at the planning phases. How well were the plans executed? What types of challenges were faced and how were they resolved? What are the lessons that should be used to improve future plans? Analysis of unsafe acts, unsafe conditions and non-compliances to approved plans and procedures will be reviewed and necessary actions to prevent recurrence will be noted. These will be used in development of future turnaround safety plans and pre-execution safety workshops. The safety performance of each of the contractors should also be analyzed, reviewed and kept in a database for reference in future bids. The product of the review meeting should be series of SMART (Specific, Measurable, Achievable, Realist and Time-bound) action items that will drive continual safety improvement in future turnaround maintenance activities.

www.ingramcontent.com/pod-product-compliance
Lightning Source LLC
Chambersburg PA
CBHW030817180526
45163CB00003B/1322